T0240052

Theoretical Thermotics

Ji-Ping Huang

Theoretical Thermotics

Transformation Thermotics and Extended
Theories for Thermal Metamaterials

 Springer

Ji-Ping Huang
Department of Physics
Fudan University
Shanghai, China

ISBN 978-981-15-2303-8   ISBN 978-981-15-2301-4   (eBook)
https://doi.org/10.1007/978-981-15-2301-4

This Springer imprint is published by the registered company Springer Nature Singapore Pte Ltd.
The registered company address is: 152 Beach Road, #21-01/04 Gateway East, Singapore 189721, Singapore

Preface

Metamaterial Physics Deserves a Nobel Prize

During the release of 2019 Nobel Prize in Physics, I was finalizing the book. This reminds me to think about an interesting (or tongue-in-cheek) problem in order to attract the reader: Does metamaterial physics deserve to be issued a Nobel prize? Absolutely, my answer is "YES". See Fig. 1. Since the seminal article by V. G. Veselago (June 13, 1929—September 15, 2018) in 1968 and especially the two other seminal articles by J. B. Pendry and coauthors in 1996 and 1999, the field of metamaterial physics has grown vigorously until today. With the aid of the

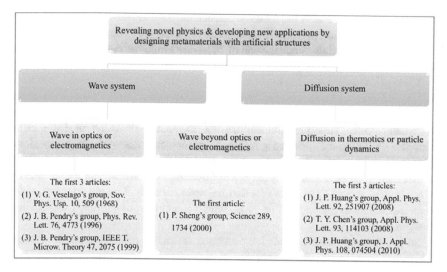

Fig. 1 A large number of novel physics and applications have arisen from metamaterials with artificial structures for wave systems and diffusion systems since 1968 and 2008, respectively. Both waves and diffusion are two important methods for transferring energy. See also Appendix: Brief History of the First Ten Years of Thermal Metamaterials

concept of metamaterial, many fundamental physics have been discovered in various branches of physics, ranging from optics/electromagnetics to elasticity/acoustics/mechanics/⋯ for wave systems, and from thermotics to particle dynamics for diffusion systems. As a result, various kinds of metamaterials were theoretically designed and experimentally fabricated in such branches. This book focuses on the branch of thermotics, namely, thermal metamaterials. The phrase "thermal metamaterial" was first adopted in Ref. [1] to name thermal cloaks (shields) and relevant devices designed by using transformation thermotics for heat conduction (diffusion) studied in the five references [2–6]. Owing to the existence of three ways of heat transfer (i.e., conduction, convection, and radiation), nowadays the connotation of "thermal metamaterial" has naturally been extended to include metamaterials for controlling heat convection and radiation. Incidentally, in this book, thermal metamaterials also contain some thermal metadevices (whose novel functions are realized mainly because of specific geometric structures), to comply with the common usage in the literature.

Thermal Metamaterial: Past, Present, and Future

In 2008, my group and Chen's group predicted the concept of novel thermal phenomena including thermal cloaking one after another [2, 3]. At the early stage (loosely speaking, before 2014) of thermal metamaterials, many experiments have been conducted to demonstrate the phenomenon of thermal cloaking under various conditions, see Refs. [5–9]. Accordingly, this field received plenty of popular attention [10–12] (see also https://www.sciencemag.org/news/2012/05/heat-trickery-paves-way-thermal-computers). These impacts attracted me to come back to the field of thermal metamaterials. Since the end of 2014, my group has completely returned to this field. So far, we have published dozens of articles.

Thermal metamaterials mean those materials or devices with artificial structures that can be used to control heat conduction, convection, and radiation in novel manners. In this case, geometric structure (rather than physical property) plays a dominating role. This fact makes thermal metamaterials different from other materials including thermoelectric materials, pyroelectric materials, magnetocaloric materials, and photothermal conversion materials; for the latter, physical property (rather than geometric structure) plays a dominating role instead. For a brief history of the first 10 years (2008–2018) for thermal metamaterials, I would refer the reader to the Appendix at the end of this book, which is a celebration article I was invited to write.

So far, thermal metamaterials have aroused enormous research interests, as also evidenced by Google search that shows the search of "thermal metamaterials" occupies 29.6% of all kinds of "metamaterials" as of August 13, 2019.

To celebrate the fruitful progress of thermal metamaterials and to prepare for the future challenges, I launched and chaired a National Conference on Thermodynamics and Thermal Metamaterials on July 18–19, 2019, in Fudan

Fig. 2 Group photo: 2019 National Conference on Thermodynamics and Thermal Metamaterials, held on July 18–19, 2019, in Fudan University, Shanghai, China

University, Shanghai, China. About 40 participants attended the first national conference, see Fig. 2. Due to the success of this national conference, I plan to not only continue the holding of the national conference, but also launch an international conference on the same topic starting from 2020.

Useful Theoretical Physics and Useful Theoretical Thermotics

To design thermal metamaterials in the literature, analytic theories have been extensively developed with a special focus on transformation thermotics. In this book, I would prefer to call the transformation thermotics and its extended theories together as "theoretical thermotics", with an attempt to contribute them to the discipline of "theoretical physics (statistical physics)" that is developing very well in China due to the efforts of many good researchers. This name could also remind the colleagues and latecomers to figure out the microscopic mechanisms for "theoretical thermotics" (that, after all, mainly describes macroscopic thermal theories for the time being), rather than to satisfy with the existing macroscopic theories; in this direction, Chap. 8 already gives a good example. Consequently, the name "theoretical thermotics" looks more suitable than other candidates like "structural thermotics" or "artificial thermotics" (the latter have been kindly suggested to me by some friends of mine).

In a word, theoretical thermotics describes the theory of transformation thermotics and its extended theories for the *active control* of macroscopic thermal properties of artificial systems, namely, metamaterials with artificial structures. Thus, theoretical thermotics is in sharp contrast to classical thermodynamics, which mainly comprises the four thermodynamic laws with a particular emphasis on the *passive description* of macroscopic thermal properties of natural systems. Incidentally, because the transformation method in transformation thermotics and theoretical thermotics is not intended to transform (or actually cannot transform) the four thermodynamic laws in thermodynamics, for the sake of clarity I choose the wording "thermotics" instead of "thermodynamics" for naming transformation thermotics or theoretical thermotics.

Clearly, theoretical thermotics can help to design thermal metamaterials, which are further useful for engineering techniques and applications [13], say, for designing standard printed circuit board [14, 15], daytime radiative cooling [16], and so on. This book focuses on fundamental theories, rather than engineering techniques and applications, and it introduces 18 theories including 7 general theories and 11 special theories.

Acknowledgement and Some Additional Notes

The main content of this book mainly comes from the articles published by my group. The current members of my group are Mr. C. R. Jiang, Mr. G. L. Dai, Mr. J. Wang, Ms. S. Yang, Mr. B. Y. Tian, Mr. L. J. Xu, Mr. F. B. Yang, Dr. B. Wang, Mr. P. Jin, and Mr. C. Q. Wang, and they also helped to improve the English of this book. Some former group members working in the field include Dr. C. Z. Fan, Dr. Y. Gao, Mr. J. Y. Li, Dr. X. Y. Shen, and Dr. Y. Li. I must thank all the current and past members for their fruitful contributions, which make this book possible.

In particular, together with me, Mr. G. L. Dai wrote Chaps. 2, 3, 6, and 8, Mr. L. J. Xu and Mr. G. L. Dai wrote Chaps. 4 and 5, and Ms. S. Yang translated the Appendix from Chinese to English. Also, Mr. G. L. Dai, Mr. L. J. Xu, and Ms. S. Yang helped to prepare the section "Exercises and Solutions" after each chapter. The reason why we add "Exercises and Solutions" is that we hope this book could be not only a monograph for experts to read, but also a textbook for newcomers to practice (so that he/she could engage in this new field as soon as possible). Incidentally, in order to facilitate reading, each chapter in the book has its own symbols. In this sense, to read the book, the reader can start with any chapter of the book (especially, Chaps. 6–19).

In the last years, I have had many invaluable opportunities to present our research progress on theoretical thermotics and thermal metamaterials to many top professors (in no particular order): Prof. Chang-Pu Sun, Prof. Rong-Gen Cai, Prof. Min-Xing Luo, Prof. Zhong-Xian Zhao, Prof. Qi Ou-Yang, Prof. Yu-Gang Ma, Prof. Xiao-Ping Ou-Yang, Prof. Ji Zhou, Prof. Zhong-Can Ou-Yang, Prof. Ding-Yu Xing, Prof. Shi-Ning Zhu, Prof. Yi-Peng Jing, Prof. Hong-Xing Xu, Prof. Bao-Wen Li, Prof. Qi-Kun Xue, Prof. Xin-Cheng Xie, Prof. Rui-Bao Tao, Prof. Xin-Gao Gong, Prof. Lu Yu, Prof. Tao Xiang, Prof. Xian-Hui Chen, Prof. Xian-Gang Luo, Prof. Tie-Jun Cui, Prof. Wei Wang, Prof. Zheng-You Liu,

Prof. Hai-Qing Lin, Prof. Ke-Qing Xia, Prof. Mu Wang, Prof. Xiao-Peng Zhao, Prof. Shu-Xin Bai, Prof. Hong Zhao, Prof. Bo Zheng, Prof. Yuan-Ning Gao, Prof. Dong-Lai Feng, and Prof. Hai-Ping Fang. Here, I want to thank all of them from the bottom of my heart for their critical comments, inspiring encouragement, and unceasing support.

I am also indebted to my family members, especially two daughters (Ji-Yan Huang with a nickname of Qian-Qian and Ji-Yang Huang with a nickname of Yue-Yue), for bringing me happiness.

Last but not least, I acknowledge the financial support by the National Natural Science Foundation of China under Grant No. 11725521.

Shanghai, China   Ji-Ping Huang
October 2019

Bibliography

1. Maldovan, M.: Sound and heat revolutions in phononics. Nature **503**, 209–217 (2013)
2. Fan, C.Z., Gao, Y., Huang, J.P.: Shaped graded materials with an apparent negative thermal conductivity. Appl. Phys. Lett. **92**, 251907 (2008)
3. Chen, T.Y., Weng, C.N., Chen, J.S.: Cloak for curvilinearly anisotropic media in conduction. Appl. Phys. Lett. **93**, 114103 (2008)
4. Guenneau, S., Amra, C., Veynante, D.: Transformation thermodynamics: cloaking and concentrating heat flux. Opt. Express **20**, 8207–8218 (2012)
5. Narayana, S., Sato, Y.: Heat flux manipulation with engineered thermal materials. Phys. Rev. Lett. **108**, 214303 (2012)
6. Schittny, R., Kadic, M., Guenneau, S., Wegener, M.: Experiments on transformation thermodynamics: molding the flow of heat. Phys. Rev. Lett. **110**, 195901 (2013)
7. Han, T.C., Bai, X., Gao, D.L., Thong, J.T.L., Li, B.W., Qiu, C.-W.: Experimental demonstration of a bilayer thermal cloak. Phys. Rev. Lett. **112**, 054302 (2014)
8. Xu, H.Y., Shi, X.H., Gao, F., Sun, H.D., Zhang, B.L.: Ultrathin three-dimensional thermal cloak. Phys. Rev. Lett. **112**, 054301 (2014)
9. Ma, Y.G., Liu, Y.C., Raza, M., Wang, Y.D., He, S.L.: Experimental demonstration of a multiphysics cloak: Manipulating heat flux and electric current simultaneously. Phys. Rev. Lett. **113**, 205501 (2014)
10. Leonhardt, U.: Cloaking of heat. Nature **498**, 440–441 (2013)
11. Wegener, M.: Metamaterials beyond optics. Science **342**, 939–940 (2013)
12. Ball, P.: Against the flow. Nature Mater. **11**, 566–566 (2012)
13. Huang, J.P.: Technologies for Controlling Thermal Energy: Design, Simulation and Experiment based on Thermal Metamaterial Theories including Transformation Thermotics (in Chinese). Higher Education Press, Beijing (2020)
14. Dede, E.M., Schmalenberg, P., Nomura, T., Ishigaki, M.: Design of anisotropic thermal conductivity in multilayer printed circuit boards. IEEE Trans. Compon. Packag. Manuf. Technol. **5**, 1763–1774 (2015)
15. Dede, E.M., Zhou, F., Schmalenberg, P., Nomura, T.: Thermal metamaterials for heat flow control in electronics. J. Electron. Packag. **140**, 010904 (2018)
16. Zhai, Y., Ma, Y.G., David, S.N., Zhao, D.L., Lou, R.N., Tan, G., Yang, R.G., Yin, X.B.: Scalable-manufactured randomized glass-polymer hybrid metamaterial for daytime radiative cooling. Science **355**, 1062–1066 (2017)

Contents

Chapter 1
Introduction

Abstract Classical thermodynamics pays a special attention to the passive description of macroscopic heat phenomena of natural systems with the theoretical framework of the four thermodynamic laws. In contrast, theoretical thermotics, introduced in this book, allows one to achieve the active control of macroscopic heat phenomena of artificial systems with the theoretical framework of transformation thermotics and extended theories. As a result, thermal metamaterials can be theoretically designed at will, which have abundant application values. Thus, a hot field comes to appear.

Keywords Thermodynamics · Theoretical thermotics · Passive description · Active control · Transformation thermotics · Thermal metamaterials

1.1 Thermodynamics Versus Theoretical Thermotics

1.1.1 Thermodynamics Concentrating on a Passive Description of Macroscopic Heat Phenomena of Natural Systems

The framework of thermodynamics is composed of the four laws of thermodynamics. Let us take the second law of thermodynamics as an example, which states "the total entropy of an isolated system can never decrease over time". The statement indicates an intrinsic property of isolated systems, and this property can not be changed by humans at all. Thus, we would say that classical thermodynamics pays a special attention on the passive description of macroscopic heat phenomena of natural systems.

© Springer Nature Singapore Pte Ltd. 2020
J.-P. Huang, *Theoretical Thermotics*,
https://doi.org/10.1007/978-981-15-2301-4_1

1.1.2 Theoretical Thermotics Concentrating on an Active Control of Macroscopic Heat Phenomena of Artificial Systems

The above-mentioned passive description of macroscopic heat phenomena means that humans can not break the four laws, but only obey them. In this regard, if one can control heat flow at will, this control would be definitely useful for human life. This is just the goal of theoretical thermotics. Certainly, the four laws of thermodynamics also work for theoretical thermotics, but we try to establish and develop different kinds of theories to manipulate and control the flow of heat purposefully. Consequently, we achieve the active control of macroscopic heat phenomena of artificial systems.

1.2 Two Features of Theoretical Thermotics

1.2.1 Theoretical Framework: Transformation Thermotics and Extended Theories

In theoretical framework, we establish and develop the theory of transformation thermotics and its extended theories (all are analytical theories). Such theories allow us to design artificial systems or structures (thermal metamaterials), in order to control heat transfer arbitrarily.

1.2.2 Application Value: Design Thermal Metamaterials for Macroscopic Heat-Flow Control

Thermal metamaterials pave a new way to control the transfer of heat (conduction, convection, and radiation). For the sake of comprehensiveness, below we present more relevant backgrounds and details according to Ref. [1].

With the advent of energy crisis, energy sources like coal, oil and natural gas are becoming less and less. However, more and more low-grade heat energy is produced and wasted due to various reasons including inefficient utilization. Therefore, how to efficiently control the flow of heat energy becomes particularly important.

Heat transfer at microscopic scale has been deeply explored by many scholars, such as Refs. [2–8], which have helped to develop the field significantly. For the existing research at microscopic scale, a delicate review has been made by [5]. In contrast, the topic of this chapter and this book is mainly on theories and experiments for controlling heat transfer at macroscopic scale. Certainly, traditional Fourier's law (bridging heat flux and temperature gradient in a material), established by Joseph

Fourier in his treatise "Théorie analytique de la chaleur" (1822), can be seen as the first quantitative theory for studying heat conduction at macroscopic scale. After 1822, about two hundred years have witnessed much more developments, such as, applying effective medium theories from optics/electromagnetics [9, 10] to thermotics due to the mathematical similarity between dielectric permittivities and thermal conductivities. Such theories have been reviewed by many researchers including [11, 12]. Meanwhile, many other macroscopic methods have also been proposed to study heat transfer, such as phonon hydrodynamics models [13, 14], the dual-phase-lag model [15, 16], the ballistic-diffusive model [17, 18], and so on. Such methods can be referred to a comprehensive review by Guo and Wang [19].

Starting from ten years ago, researchers started to develop new theories for controlling macroscopic heat transfer again. Reference [20] first introduced the theory of coordinate transformation from optics/electromagnetics [21, 22] to thermotics (steady-state heat conduction), and predicted the concept of "thermal cloak", which helps to guide the flow of heat around an object as if the object does not exist. Such a thermal cloak has potential applications in thermal protection, misleading infrared detection, and heat preservation/dissipation. As a result, a new direction forms, which is called "transformation thermotics" (or equivalently "transformation thermodynamics" as occasionally used by some other researchers) in the literature.

With the establishment of transformation thermotics and extended theories, there comes a research upsurge of achieving novel thermal transport phenomena via designing artificial structures or devices. The theoretical proposals of thermal cloaks [20, 23–27] have further motivated experimental demonstrations [28–32] and popular attention [33–35] (see also http://www.sciencemag.org/news/2012/05/heat-trickery-paves-way-thermal-computers). In this book, we call transformation thermotics and extended theories as theoretical thermotics, which has been explained in Part III of Preface.

The so-called "thermal metamaterial" was first adopted by [36] to name thermal cloaks (shields) and relevant devices designed by using transformation thermotics in the five references [20, 23, 26, 28, 29], thus causing the formation of the direction of thermal metamaterials. Incidentally, the phrase "thermal metamaterial" was originally used for thermal conduction only [36], but its connotation has been significantly extended afterwards. So far, thermal metamaterials also cover those artificial structural materials for controlling thermal convection [37–39] and radiation [40–43] with novel properties. Nowadays, as defined by [44], "thermal metamaterials are materials composed of engineered, microscopic structures that exhibit unique thermal performance characteristics based primarily on their physical structures and patterning, rather than just their chemical composition or bulk material properties".

In our eyes, the existing materials for macroscopic heat control can be generally classified into two types. One is based on physical properties, such as thermoelectric materials, pyroelectric materials, magnetocaloric materials, photo thermal conversion materials, etc. The other is based upon geometric structures rather than physical properties (namely, geometric structures play a more important role than materials' physical properties). Among geometric structures, (normal) structural materials can be used to realize normal control of heat flow, but thermal metamaterials can be utilized to achieve novel controls. So far, the field of thermal metamaterials has aroused enormous research interests, as also evidenced by Google search that shows the search of "thermal metamaterials" occupies 29.6% of all kinds of "metamaterials" as of August 13, 2019.

References

1. Huang, J.P.: Thermal metamaterial: geometric structure, working mechanism, and novel function. Prog. Phys. **38**, 219 (2018)
2. Li, B.W., Wang, L., Casati, G.: Thermal diode: rectification of heat flux. Phys. Rev. Lett. **93**, 184301 (2004)
3. Wang, L., Li, B.W.: Thermal logic gates: computation with phonons. Phys. Rev. Lett. **99**, 177208 (2007)
4. Wang, L., Li, B.W.: Thermal memory: a storage of phononic information. Phys. Rev. Lett. **101**, 267203 (2008)
5. Li, N.B., Ren, J., Wang, L., Zhang, G., Hänggi, P., Li, B.W.: Phononics: manipulating heat flow with electronic analogs and beyond. Rev. Mod. Phys. **84**, 1045–1066 (2012)
6. Ben-Abdallah, P., Biehs, S.-A.: Near-field thermal transistor. Phys. Rev. Lett. **112**, 044301 (2014)
7. Kubytskyi, V., Biehs, S.-A., Ben-Abdallah, P.: Radiative bistability and thermal memory. Phys. Rev. Lett. **113**, 074301 (2014)
8. Huang, C.L., Lin, Z.Z., Luo, D.C., Huang, Z.: Electronic thermal conductivity of 2-dimensional circular-pore metallic nanoporous materials. Phys. Lett. A **380**, 3103–3106 (2016)
9. Garnett, J.C.M.: Colours in metal glasses and in metallic films. Philos. Trans. R. Soc. London Ser. A **203**, 385 (1904)
10. Bruggeman, D.A.G.: Berechnung verschiedener physikalischer Konstanten von heterogenen substanzen. I. Dielektrizitätskonstanten und Leitfähigkeiten der Mischkörper aus isotropen Substanzen (Calculation of different physical constants of heterogeneous substances. I. Dielectricity and conductivity of mixtures of isotropic substances). Annalen der Physik **24**, 636–664 (1935)
11. Bergman, D.J., Stroud, D.: Physical properties of macroscopically inhomogeneous media. Solid State Phys. **46**, 147–269 (1992)
12. Huang, J.P., Yu, K.W.: Enhanced nonlinear optical responses of materials: composite effects. Phys. Rep. **431**, 87–172 (2006)
13. Guyer, R.A., Krumhansl, J.A.: Solution of the linearized phonon Boltzmann equation. Phys. Rev. **148**, 766–778 (1966)
14. Guyer, R.A., Krumhansl, J.A.: Thermal conductivity, second sound, and phonon hydrodynamic phenomena in nonmetallic crystals. Phys. Rev. **148**, 778–788 (1966)
15. Tzou, D.Y.: The generalized lagging response in small-scale and high-rate heating. Int. J. Heat Mass Transfer **38**, 3231–3240 (1995)
16. Tzou, D.Y.: A unified field approach for heat conduction from macro- to micro-scales. J. Heat Transfer **117**, 8–16 (1995)
17. Chen, G.: Ballistic-diffusive heat-conduction equations. Phys. Rev. Lett. **86**, 2297 (2001)
18. Chen, G.: Ballistic-diffusive equations for transient heat conduction from nano to macroscales. J. Heat Transfer **124**, 320–328 (2002)
19. Guo, Y.Y., Wang, M.R.: Phonon hydrodynamics and its applications in nanoscale heat transport. Phys. Rep. **595**, 1–44 (2015)
20. Fan, C.Z., Gao, Y., Huang, J.P.: Shaped graded materials with an apparent negative thermal conductivity. Appl. Phys. Lett. **92**, 251907 (2008)

21. Pendry, J.B., Schurig, D., Smith, D.R.: Controlling electromagnetic fields. Science **312**, 1780–1782 (2006)
22. Leonhardt, U.: Optical conformal mapping. Science **312**, 1777–1780 (2006)
23. Chen, T.Y., Weng, C.N., Chen, J.S.: Cloak for curvilinearly anisotropic media in conduction. Appl. Phys. Lett. **93**, 114103 (2008)
24. Li, J.Y., Gao, Y., Huang, J.P.: A bifunctional cloak using transformation media. J. Appl. Phys. **108**, 074504 (2010)
25. Yu, G.X., Lin, Y.F., Zhang, G.Q., Yu, Z., Yu, L.L., Su, J.: Design of square-shaped heat flux cloaks and concentrators using method of coordinate transformation. Front. Phys. China **6**, 70–73 (2011)
26. Guenneau, S., Amra, C., Veynante, D.: Transformation thermodynamics: cloaking and concentrating heat flux. Opt. Express **20**, 8207–8218 (2012)
27. Han, T.C., Yuan, T., Li, B.W., Qiu, C.-W.: Homogeneous thermal cloak with constant conductivity and tunable heat localization. Sci. Rep. **3**, 1593 (2013)
28. Narayana, S., Sato, Y.: Heat flux manipulation with engineered thermal materials. Phys. Rev. Lett. **108**, 214303 (2012)
29. Schittny, R., Kadic, M., Guenneau, S., Wegener, M.: Experiments on transformation thermodynamics: molding the flow of heat. Phys. Rev. Lett. **110**, 195901 (2013)
30. Han, T.C., Bai, X., Gao, D.L., Thong, J.T.L., Li, B.W., Qiu, C.-W.: Experimental demonstration of a bilayer thermal cloak. Phys. Rev. Lett. **112**, 054302 (2014)
31. Xu, H.Y., Shi, X.H., Gao, F., Sun, H.D., Zhang, B.L.: Ultrathin three-dimensional thermal cloak. Phys. Rev. Lett. **112**, 054301 (2014)
32. Ma, Y.G., Liu, Y.C., Raza, M., Wang, Y.D., He, S.L.: Experimental demonstration of a multiphysics cloak: manipulating heat flux and electric current simultaneously. Phys. Rev. Lett. **113**, 205501 (2014)
33. Leonhardt, U.: Cloaking of heat. Nature **498**, 440–441 (2013)
34. Wegener, M.: Metamaterials beyond optics. Science **342**, 939–940 (2013)
35. Ball, P.: Against the flow. Nature Mater. **11**, 566 (2012)
36. Maldovan, M.: Sound and heat revolutions in phononics. Nature **503**, 209–217 (2013)
37. Guenneau, S., Petiteau, D., Zerrad, M., Amra, C., Puvirajesinghe, T.: Transformed Fourier and Fick equations for the control of heat and mass diffusion. AIP Adv. **5**, 053404 (2015)
38. Dai, G.L., Shang, J., Huang, J.P.: Theory of transformation thermal convection for creeping flow in porous media: cloaking, concentrating, and camouflage. Phys. Rev. E **97**, 022129 (2018)
39. Dai, G.L., Huang, J.P.: A transient regime for transforming thermal convection: cloaking, concentrating and rotating creeping flow and heat flux. J. Appl. Phys. **124**, 235103 (2018)
40. Raman, A.P., Anoma, M.A., Zhu, L.X., Rephaeli, E., Fan, S.H.: Passive radiative cooling below ambient air temperature under direct sunlight. Nature **515**, 540–544 (2014)
41. Shi, N.N., Tsai, C.C., Camino, F., Bernard, G.D., Yu, N.F., Wehner, R.: Keeping cool: enhanced optical reflection and radiative heat dissipation in Saharan silver ants. Science **349**, 298–301 (2015)
42. Zhai, Y., Ma, Y.G., David, S.N., Zhao, D.L., Lou, R.N., Tan, G., Yang, R.G., Yin, X.B.: Scalable-manufactured randomized glass-polymer hybrid metamaterial for daytime radiative cooling. Science **355**, 1062–1066 (2017)
43. Xu, L.J., Huang, J.P.: Metamaterials for manipulating thermal radiation: transparency, cloak, and expander. Phys. Rev. Appl. **12**, 044048 (2019)
44. Roman, Jr., C.T., Coutu, R.A., Starman, L.A.: Thermal management and metamaterials. In: MEMS and Nanotechnology (The Society for Experimental Mechanics, Inc., edited by T. Proulx), vol. 2, pp. 107–113 (2011)

Part I
General Theories

Chapter 2
Transformation Thermotics for Thermal Conduction

Abstract This chapter describes the theory of transformation thermotics for thermal conduction. We begin with the relationship between coordinate transformation and geometric transformation and then give some basic tools of tensor analysis. Based on Fourier's law for heat conduction, we show how the form-invariance of an equation under arbitrary coordinate transformation can result in a new technique to manipulate temperature field and heat flux. As a model application, we design a thermal cloak to show how transformation thermotics works.

Keywords Transformation thermotics · Coordinate transformation · Geometric transformation · Form invariance · Heat conduction

2.1 Opening Remarks

"Transformation thermotics is based on the form-invariance of the governing equations of heat transfer under coordinate transformations. It engineers thermal properties of materials like thermal conductivity, to modulate the heat flux in novel manners like cloaking, concentrating and rotating."

We can find similar descriptions about transformation thermotics [1, 2] in the literature today. If one is not familiar with transformation theory on thermotics, optics or acoustics, he/she might be puzzled by some concepts like "form-invariance under coordinate transformations" and why this invariance can be used for heat management. Here, we shall firstly talk about the motivation of transforming theory and introduce some basic concepts.

Suppose light is traveling on a uniform plane and the trace of movement is a straight line. Now one wants to let the light move on a curve, a simple idea is just to bend the plane and then he/she may expect the light is bent accordingly. However, is this enough and how to bend the space like bending a paper? Luckily, we have been told in general relativity that the change of energy-momentum tensor can bend the space so we can have a more general guess here that if one wants to manipulate some physical fields as if the space is changed, he/she can change some important properties of the space or the material on it, for example, the thermal conductivity tensor.

© Springer Nature Singapore Pte Ltd. 2020
J.-P. Huang, *Theoretical Thermotics*,
https://doi.org/10.1007/978-981-15-2301-4_2

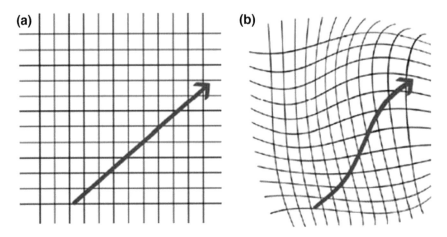

Fig. 2.1 Schematic diagram showing how transformation works, **a** is the original coordinate system shown by black uniform grids, and the blue arrow represents a straight physical field, **b** is the new coordinate system shown by uneven grids, which can also be seen as a twisted space so the blue arrow is curved

Now one may ask: "Where is the coordinate transformation? You seem be talking about geometric transformation when mentioning bending the light. What's more, why can this idea work for heat transfer?" To answer these questions, we should introduce transformation theory which tells how to change space or material properties based on coordinate transformation to achieve the desired effect as the fields change under geometric transformation; see Fig. 2.1. Also, we shall discuss the condition when transformation theory is valid.

2.2 Coordinate Transformation and Geometric Transformation

Let us start from the relationship between coordinate transformation and geometric transformation. For clarity, we have to talk about some basic knowledge on tensor analysis. Using Cartesian coordinate system in three-dimensional Euclidean space $\mathbb{E}^3$, a vector r with coordinates (x, y, z) can be written as

$$r = x\boldsymbol{i} + y\boldsymbol{j} + z\boldsymbol{k} \tag{2.1}$$

where $\{\boldsymbol{i}, \boldsymbol{j}, \boldsymbol{k}\}$ is the standard orthogonal basis of Cartesian coordinate system. Consider a mapping $f : \mathbb{E}^3 \to \mathbb{E}^3$, satisfying

$$f(r) = (2x)\boldsymbol{i} + (2y)\boldsymbol{j} + (2z)\boldsymbol{k}. \tag{2.2}$$

It can be easily checked that f is a bijection or one-to-one correspondence on $\mathbb{E}^3$. The meaning of f is that the length of each vector doubles in $\mathbb{E}^3$ while the direction keeps unchanged. For a unit-ball in $\mathbb{E}^3$, its volume becomes 8 times under f. This is a simple example of geometric transformation, which changes the vector r.

Naturally, we have another bijection $\hat{f} : \mathbb{R}^3 \to \mathbb{R}^3$, satisfying

$$\hat{f}((x, y, z)) = (2x, 2y, 2z). \tag{2.3}$$

If we use a new set of basis $\{g_u, g_v, g_w\} = \{i/2, j/2, k/2\}$, we can see $\hat{f}$ just gives the new coordinates under this basis,

$$(x, y, z) \cdot (i, j, k)' = \hat{f}((x, y, z)) \cdot (g_u, g_v, g_w)'. \tag{2.4}$$

Here we should point out that a set of vectors $\{g_u, g_v, g_w\}$ can be a basis in $\mathbb{E}^3$ if and only if they are linearly unrelated (the 3 vectors are not in the same plane). In other words, orthogonality and normality are unnecessary. $\{g_u, g_v, g_w\}$ is also called covariant basis. In tensor analysis, contravariant basis $\{g^u, g^v, g^w\}$ is another set of vectors satisfying

$$g_u \cdot g^v = \delta_{uv}, \tag{2.5}$$

where δ_{uv} is the Kronecker delta

$$\delta_{uv} = \begin{cases} 0 & \text{if } u \neq v, \\ 1 & \text{if } u = v. \end{cases} \tag{2.6}$$

It is obvious to see the existence of this contravariant basis and we may decompose the vector r as

$$r = x^u g_u + x^v g_v + x^w g_w = x_u g^u + x_v g^v + x_w g^w, \tag{2.7}$$

or by using Einstein summation convention, we can simplify it as

$$r = x^u g_u = x_u g^u. \tag{2.8}$$

Here $\{x^u, x^v, x^w\}$ is also known as contravariant components and $\{x_u, x_v, x_w\}$ is called covariant components, which can be obtained by

$$x^u = r \cdot g^u, \quad x_u = r \cdot g_u. \tag{2.9}$$

In Cartesian coordinate systems, both covariant basis and contravariant basis are $\{i, j, k\}$, so covariant and contravariant components are also the same.

To sum up, coordinate transformation means choosing a different basis while the vector r itself is not changed. In fact, invariance under coordinate transformation is a necessary condition for vectors.

So far, we can see geometric transformation and coordinate transformation are two different concepts. However, it can be observed that the mapping f in geometric transformation and mapping $\hat{f}$ in coordinate transformation have some relationships. Mapping f can naturally induce mapping $\hat{f}$ and vice versa. They both change the coordinates (and thus length) of a vector: f changes the vector itself while $\hat{f}$ changes the measure of space instead. So, we can take f and $\hat{f}$ as the same if we only care about the mathematical forms of new coordinates after the mappings, although they have different physical explanations indeed.

For most curvilinear coordinate systems, $\{g_u, g_v, g_w\}$ is not a set of constant vectors and can vary with the elements in $\mathbb{E}^3$. Unless otherwise stated in this chapter, indices u, v, w are used for general (curvilinear) coordinate systems while i, j, k for Cartesian coordinate systems. For example, in spherical coordinate systems, we have $r = rg_u + \theta g_v + \varphi g_w$ where

$$
\begin{aligned}
g_u &= \sin\theta\cos\varphi\, i + \sin\theta\sin\varphi\, j + \cos\theta k, \\
g_v &= r(\cos\theta\cos\varphi\, i + \cos\theta\sin\varphi\, j - \sin\theta k), \\
g_w &= r\sin\theta(-\sin\varphi\, i + \cos\varphi\, j).
\end{aligned}
\tag{2.10}
$$

In addition, we can see only g_u is a unit vector. Here $\{g_u, g_v, g_w\}$ is also called local covariant basis and we shall give the derivation for general cases below. Let (x^u, x^v, x^w) denote the coordinates for a vector in a curvilinear coordinate system which has the relationship with Cartesian coordinates as

$$
\begin{aligned}
x^u &= x^u(x, y, z), \\
x^v &= x^v(x, y, z), \\
x^w &= x^w(x, y, z).
\end{aligned}
\tag{2.11}
$$

To ensure (x^u, x^v, x^w) can be a curvilinear coordinate, the map $\hat{f} : (x, y, z) \rightarrow (x^u, x^v, x^w)$ should be a smooth bijection, which is equivalent to the condition

$$
\det \mathbf{J} =
\begin{vmatrix}
\dfrac{\partial x^u}{\partial x} & \dfrac{\partial x^u}{\partial y} & \dfrac{\partial x^u}{\partial z} \\[2mm]
\dfrac{\partial x^v}{\partial x} & \dfrac{\partial x^v}{\partial y} & \dfrac{\partial x^v}{\partial z} \\[2mm]
\dfrac{\partial x^w}{\partial x} & \dfrac{\partial x^w}{\partial y} & \dfrac{\partial x^w}{\partial z}
\end{vmatrix}
\neq 0, \quad
\det \mathbf{J}^{-1} =
\begin{vmatrix}
\dfrac{\partial x}{\partial x^u} & \dfrac{\partial x}{\partial x^v} & \dfrac{\partial x}{\partial x^w} \\[2mm]
\dfrac{\partial y}{\partial x^u} & \dfrac{\partial y}{\partial x^v} & \dfrac{\partial y}{\partial x^w} \\[2mm]
\dfrac{\partial z}{\partial x^u} & \dfrac{\partial z}{\partial x^v} & \dfrac{\partial z}{\partial x^w}
\end{vmatrix}
\neq 0,
\tag{2.12}
$$

where $\mathbf{J}$ is the Jacobian matrix (we use a different font to distinguish between tensors) from coordinate (x, y, z) to (x^u, x^v, x^w). Here, the domain (for (x, y, z)) and the range (for (x^u, x^v, x^w)) of $\hat{f}$ are both $\mathbb{R}^3$.

Since we want to obtain the local basis for vector r with coordinate (x^u, x^v, x^w), we write the line element for an infinitesimal displacement from r to $r + dr$,

$$
dr = \frac{\partial r}{\partial x^u} dx^u + \frac{\partial r}{\partial x^v} dx^v + \frac{\partial r}{\partial x^w} dx^w.
\tag{2.13}
$$

On the other hand, for vector $d\boldsymbol{r}$, its coordinate is set as (dx^u, dx^v, dx^w), meaning

$$d\boldsymbol{r} = \boldsymbol{g}_u dx^u + \boldsymbol{g}_v dx^v + \boldsymbol{g}_w dx^w. \tag{2.14}$$

So the local covariant basis is just

$$\boldsymbol{g}_u = \frac{\partial \boldsymbol{r}}{\partial x^u}, \quad \boldsymbol{g}_v = \frac{\partial \boldsymbol{r}}{\partial x^v}, \quad \boldsymbol{g}_w = \frac{\partial \boldsymbol{r}}{\partial x^w}. \tag{2.15}$$

It is clear that $\{\boldsymbol{g}_u, \boldsymbol{g}_v, \boldsymbol{g}_w\}$ points out the directions in which (u, v, w) increases. Finally we have

$$\begin{aligned}
\boldsymbol{g}_u &= \frac{\partial x}{\partial x^u}\boldsymbol{i} + \frac{\partial y}{\partial x^u}\boldsymbol{j} + \frac{\partial z}{\partial x^u}\boldsymbol{k}, \\
\boldsymbol{g}_v &= \frac{\partial x}{\partial x^v}\boldsymbol{i} + \frac{\partial y}{\partial x^v}\boldsymbol{j} + \frac{\partial z}{\partial x^v}\boldsymbol{k}, \\
\boldsymbol{g}_w &= \frac{\partial x}{\partial x^w}\boldsymbol{i} + \frac{\partial y}{\partial x^w}\boldsymbol{j} + \frac{\partial z}{\partial x^w}\boldsymbol{k}.
\end{aligned} \tag{2.16}$$

This is a very convenient choice of the basis and we can use other basis. With local basis, we can introduce metric tensor $\boldsymbol{G}$, whose covariant components are defined as

$$g_{uv} = \boldsymbol{g}_u \cdot \boldsymbol{g}_v. \tag{2.17}$$

Then we can use the form of tensor product $\otimes$ (the Cartesian product) as

$$\boldsymbol{G} = g_{uv}\boldsymbol{g}^u \otimes \boldsymbol{g}^v. \tag{2.18}$$

The determinant of $[g_{uv}]$ is

$$g = |[g_{uv}]| \tag{2.19}$$

and it is a function with $\boldsymbol{r}$ or (x^u, x^v, x^w). Since we can also write

$$\boldsymbol{G} = g^{uv}\boldsymbol{g}_u \otimes \boldsymbol{g}_v = g^u_v \boldsymbol{g}_u \otimes \boldsymbol{g}^v = g^v_u \boldsymbol{g}^u \otimes \boldsymbol{g}_v, \tag{2.20}$$

we obtain

$$|[g^{uv}]| = \frac{1}{g}, \quad |[g^u_v]| = |[g^v_u]| = 1. \tag{2.21}$$

Here what we want to emphasize is that the determinant of a rank-2 tensor is different from the determinant of a matrix. In tensor analysis, the determinant of a rank-2 tensor $\boldsymbol{A}$ is

$$\det \boldsymbol{A} = |[A^u_v]| = |[A^v_u]|. \tag{2.22}$$

For metric tensor $\boldsymbol{G}$, we have

$$\det \boldsymbol{G} = 1, \tag{2.23}$$

and for any rank-2 tensor A, it is easy to prove

$$G \cdot A = A \cdot G = A. \tag{2.24}$$

In this sense, G behaves as an identity rank-2 tensor.

The derivation for local basis mentioned above requires no specific properties of Cartesian coordinate systems. As a result, we can directly write the relationship between two curvilinear coordinates (x^1, x^2, x^3), $(x^{(1)}, x^{(2)}, x^{(3)})$ and their local covariant bases $\{g_1, g_2, g_3\}$, $\{g_{(1)}, g_{(2)}, g_{(3)}\}$,

$$g_{(u)} = \frac{\partial x^v}{\partial x^{(u)}} g_v, \quad g_u = \frac{\partial x^{(v)}}{\partial x^u} g_{(v)}. \tag{2.25}$$

If we are concerned about how g_u varies with r, we may resort to Christoffel symbols of the second kind Γ_w^{vu},

$$\Gamma_{vu}^w = \frac{\partial g_u}{\partial x^v} \cdot g^w, \tag{2.26}$$

which implies

$$\frac{\partial g_u}{\partial x^v} = \Gamma_{vu}^w g_w. \tag{2.27}$$

One of the most useful conclusions for Christoffel symbols is

$$\Gamma_{vu}^v = \frac{1}{\sqrt{g}} \frac{\partial \sqrt{g}}{\partial x^u} \tag{2.28}$$

and the proof is left as an exercise (see Sect. 2.5).

In conclusion, we have introduced the transformation rules for the coordinates of vectors under coordinate transformation and geometric transformation. One important point is that the vector itself is invariant in different coordinate systems and so is the tensor. In particular, vector is a class of rank-1 tensor. However, the covariant and contravariant components of a tensor, which are coordinates for vectors, should change with the choice of coordinate systems (local basis). What's more, some fundamental operators on tensors should be taken into reconsideration. For example, the gradient operator ∇ is defined as

$$\nabla \equiv g^u \frac{\partial}{\partial x^u} \tag{2.29}$$

and the divergence operator $\nabla \cdot$ is defined as

$$\nabla \cdot \equiv g^u \cdot \frac{\partial}{\partial x^u}. \tag{2.30}$$

For any vector $\boldsymbol{r}$, we can easily write

$$\nabla \cdot \boldsymbol{r} = \boldsymbol{g}^u \cdot \frac{\partial}{\partial x^u} r^v \boldsymbol{g}_v = \frac{\partial r^u}{\partial x^u} = \nabla_u r^u, \tag{2.31}$$

where ∇_u represents the covariant derivative, and we can find both gradient operator and divergence operator are form-invariant under coordinate transformation.

2.3 Transforming Heat Conduction

In this section we are in a position to show how an equation will change under coordinate transformation and we will take the heat conduction equation as an example. Fourier's law for heat conduction is

$$\frac{\partial \rho(T)c_p(T)T}{\partial t} - \nabla \cdot [\kappa(T)\nabla T] = 0, \tag{2.32}$$

where ρ is the density, T is the temperature, c_p is the specific heat, and t is the time. Here $\kappa(T)$ is the thermal conductivity tensor, which, for the sake of generality, is a function of T [3, 4]. When writing this equation, we donnot claim which coordinate we use. The most important parameter here, $\kappa(T)$, is a rank-2 tensor and in the Cartesian coordinate system it is

$$\kappa(T) = \kappa^{ij}(T)\boldsymbol{i} \otimes \boldsymbol{j} = \kappa_{ij}(T)\boldsymbol{i} \otimes \boldsymbol{j}, \tag{2.33}$$

or in a more familiar form as matrix, it becomes

$$\left[\kappa(T)\right] = \begin{bmatrix} \kappa^{xx}(T) & \kappa^{xy}(T) & \kappa^{xz}(T) \\ \kappa^{yx}(T) & \kappa^{yy}(T) & \kappa^{yz}(T) \\ \kappa^{zx}(T) & \kappa^{zy}(T) & \kappa^{zz}(T) \end{bmatrix}. \tag{2.34}$$

The term $\kappa(T)\nabla T$ is the product of a tensor $\kappa(T)$ and a vector ∇T and it can be written more strictly as an inner product,

$$\kappa(T) \cdot \nabla T = \kappa^{uv}(T)\boldsymbol{g}_u \otimes \boldsymbol{g}_v \cdot \boldsymbol{g}^l \frac{\partial T}{\partial x^l}. \tag{2.35}$$

In the Cartesian coordinate system, Eq. (2.35) can be written as

$$\begin{bmatrix} \kappa^{xx}(T) & \kappa^{xy}(T) & \kappa^{xz}(T) \\ \kappa^{yx}(T) & \kappa^{yy}(T) & \kappa^{yz}(T) \\ \kappa^{zx}(T) & \kappa^{zy}(T) & \kappa^{zz}(T) \end{bmatrix} \begin{bmatrix} \partial_x(T) \\ \partial_y(T) \\ \partial_z(T) \end{bmatrix}. \tag{2.36}$$

Now, we may choose any curvilinear coordinate system and recall that gradient is defined as $\nabla \equiv g^u \dfrac{\partial}{\partial x^u}$, thus yielding

$$
\begin{aligned}
\nabla \cdot [\kappa(T) \cdot \nabla T] &= g^w \cdot \frac{\partial}{\partial x^w}\left[\kappa^{uv}(T)g_u \otimes g_v \cdot g^l \frac{\partial T}{\partial x^l}\right] \\
&= g^w \cdot \frac{\partial}{\partial x^w}\left[\kappa^{uv}(T)g_u \frac{\partial T}{\partial x^v}\right] \\
&= \frac{\partial \kappa^{uv}(T)}{\partial x^u}\frac{\partial T}{\partial x^v} + \frac{\partial^2 T}{\partial x^u \partial x^v}\kappa^{uv}(T) + g^w \cdot \frac{\partial g_u}{\partial x^w}\left[\kappa^{uv}(T)\frac{\partial T}{\partial x^v}\right] \\
&= \partial_u\left[\kappa^{uv}(T)\partial_v T\right] + \Gamma^w_{wu}\kappa^{uv}(T)\partial_v T \\
&= \partial_u\left[\kappa^{uv}(T)\partial_v T\right] + \frac{1}{\sqrt{g}}(\partial_u\sqrt{g})\kappa^{uv}(T)\partial_v T \\
&= \frac{1}{\sqrt{g}}\partial_u\left[\sqrt{g}\kappa^{uv}(T)\partial_v T\right].
\end{aligned}
\tag{2.37}
$$

Then we obtain

$$
\frac{\partial \rho(T)c_p(T)T}{\partial t} - \frac{1}{\sqrt{g}}\partial_u\left[\sqrt{g}\kappa^{uv}(T)\partial_v T\right] = 0.
\tag{2.38}
$$

Since $\sqrt{g}$ is independent of time, we can rewrite the heat conduction equation Eq. (2.32) by using tensor components of any curvilinear coordinate system as

$$
\frac{\partial \sqrt{g}\rho(T)c_p(T)T}{\partial t} - \partial_u\left[\sqrt{g}\kappa^{uv}(T)\partial_v T\right] = 0.
\tag{2.39}
$$

Now, we have proven the well-known form-invariance **for tensor components** of heat conduction in different coordinate systems, or in other words, under coordinate transformations. The only difference for different coordinate systems is the coefficient $\sqrt{g}$.

The next question is what this form-invariance can induce? Remember a coordinate transformation can always be related with a geometric transformation so we can see Eq. (2.39) as a result of geometric transformation. If we still use the Cartesian coordinate system after the geometric transformation, we can write a transformed conductivity tensor $\tilde{\kappa}(T) = \tilde{\kappa}^{ij}(T)i \otimes j$ and its matrix is

$$
[\tilde{\kappa}(T)] = \begin{bmatrix} \tilde{\kappa}^{xx}(T) & \tilde{\kappa}^{xy}(T) & \tilde{\kappa}^{xz}(T) \\ \tilde{\kappa}^{yx}(T) & \tilde{\kappa}^{yy}(T) & \tilde{\kappa}^{yz}(T) \\ \tilde{\kappa}^{zx}(T) & \tilde{\kappa}^{zy}(T) & \tilde{\kappa}^{zz}(T) \end{bmatrix} = \left[\sqrt{g}\kappa^{uv}(T)\right].
\tag{2.40}
$$

In addition, for transient conduction, the product of density and specific heat is also transformed as

$$
\tilde{\rho}(T)\tilde{c}_p(T) = \sqrt{g}\rho(T)c_p(T).
\tag{2.41}
$$

To calculate the transformed physical properties, we should first know $\sqrt{g}$ and tensor component κ_{uv} for the curvilinear coordinate (x^u, x^v, x^w). With Eqs. (2.12) and (2.16), we have

$$g_{u,v} = \mathbf{J}_{iu}^{-1}\mathbf{J}_{iv}^{-1} = \mathbf{J}_{ui}^{-\mathsf{T}}\mathbf{J}_{iv}^{-1}, \tag{2.42}$$

where $\mathbf{J}$ is again the Jacobian matrix for the geometric transformation or the related coordinate transformation from Cartesian coordinate (x, y, z) to the curvilinear coordinate (x^u, x^v, x^w). Hence we obtain

$$\sqrt{g} = \frac{1}{\det \mathbf{J}}. \tag{2.43}$$

Similarly, according to the form-invariance of tensor under coordinate transformation, we have

$$\kappa(T) = \kappa^{uv}(T)g_u \otimes g_v = \kappa^{ij}(T)i \otimes j, \tag{2.44}$$

thus leading to

$$\kappa^{uv}(T)\mathbf{J}_{iu}^{-1}\mathbf{J}_{vj}^{-\mathsf{T}}i \otimes j = \kappa^{ij}(T)i \otimes j. \tag{2.45}$$

This implies

$$\kappa^{uv}(T) = \mathbf{J}_{iu}\kappa^{ij}(T)\mathbf{J}_{vj}^{\mathsf{T}}. \tag{2.46}$$

So, in the Cartesian coordinate system after the geometric transformation, if we want to keep the form of Fourier's law, we should have the transformation for material as [3]

$$[\tilde{\kappa}(T)] = \frac{\mathbf{J}[\kappa(T)]\mathbf{J}^{\mathsf{T}}}{\det \mathbf{J}}, \tag{2.47}$$

and

$$\tilde{\rho}(T)\tilde{c}_p(T) = \frac{\rho(T)c_p(T)}{\det \mathbf{J}}. \tag{2.48}$$

Here we must emphasize that Eqs. (2.47) and (2.48) are new material parameters in Cartesian coordinate system and of course they can be expressed in other coordinate systems. Finally we can write the heat conduction equation in the space after an inverse geometric transformation $r \to r'$:

$$\tilde{\rho}(r')\tilde{C}(r')\frac{\partial T'(r')}{\partial t} - \nabla' \cdot [\tilde{\kappa}(r')\nabla'T'(r')] = 0. \tag{2.49}$$

The nabla symbol ∇' corresponds to the position vector r' in the transformed space. In addition, the new temperature distribution $T'(r') = T(r(r'))$. Here we neglect the parameter T' in density, specific heat and conductivity for simplicity of notes. In fact, the T in Eqs. (2.47) and (2.48) is just $T(r(r')) = T'(r')$.

In fact, an equation is a combination of scalars, vectors, tensors and operators, which can all be seen as tensors and should be invariant under coordinate transformation. As a result, an equation written without single tensor components is also invariant under coordinate transformation and this is consistent with a common sense that physical laws should not be influenced by the choice of different coordinate systems. However, the equations written by tensor components are not the same in different coordinate systems and sometimes even the form of equation can be different. If the equation written by tensor components is form-invariant, the transformation theory can be applied.

2.4 Application: Thermal Cloak

The most famous application of transformation theory is thermal cloaks. In transformation thermotics, obstacles inside a thermal cloak will not have any influence on external thermal signals (temperature distributions or heat flux). For simplicity, now we consider a two-dimensional mapping from (r, θ) to (r', θ') (see Fig. 2.2a),

$$r' = R_1 + \frac{R_2 - R_1}{R_2}r,$$

$$\theta' = \theta. \tag{2.50}$$

If we take it as a geometric transformation, it turns the circle $0 < r < R_2$ into a ring $R_1 < r' < R_2$ and the region $r' < R_1$ is "missing". The corresponding Jacobian matrix in region $R_1 < r' < R_2$ is

$$\mathbf{J} = \begin{pmatrix} \cos\theta & -r'\sin\theta \\ \sin\theta & r'\cos\theta \end{pmatrix} \begin{pmatrix} \dfrac{R_2 - R_1}{R_2} & 0 \\ 0 & 1 \end{pmatrix} \begin{pmatrix} \cos\theta & \sin\theta \\ -\dfrac{\sin\theta}{r} & \dfrac{\cos\theta}{r} \end{pmatrix}. \tag{2.51}$$

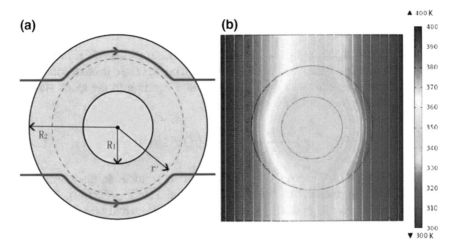

Fig. 2.2 Thermal cloak, **a** is the scheme of a thermal cloak and **b** is the temperature distribution of the thermal cloak from numerical simulations

If the original thermal conductivity is isotropic, the transformation matrix in region $R_1 < r' < R_2$ is

$$\frac{\mathbf{JJ}^\mathsf{T}}{\det \mathbf{J}} = \begin{pmatrix} \cos\theta & -r'\sin\theta \\ \sin\theta & r'\cos\theta \end{pmatrix} \begin{pmatrix} \dfrac{r'-R_1}{r'} & 0 \\ 0 & \dfrac{r'}{r'-R_1} \end{pmatrix} \begin{pmatrix} \cos\theta & -r'\sin\theta \\ \sin\theta & r'\cos\theta \end{pmatrix}^\mathsf{T}$$

$$= \begin{pmatrix} \dfrac{r'-R_1}{r'}\cos^2\theta + \dfrac{r'}{r'-R_1}\sin^2\theta & \dfrac{r'-R_1}{r'}\sin\theta\cos\theta - \dfrac{r'}{r'-R_1}\sin\theta\cos\theta \\ \dfrac{r'-R_1}{r'}\sin\theta\cos\theta - \dfrac{r'}{r'-R_1}\sin\theta\cos\theta & \dfrac{r'-R_1}{r'}\sin^2\theta + \dfrac{r'}{r'-R_1}\cos^2\theta \end{pmatrix}.$$

$$(2.52)$$

Using the transformed thermal conductivity, finite-element simulation results of temperature distribution in Fig. 2.2b show that the external temperature signals are not disturbed since the isotherms are still straight and uniform. In addition, thermal cloaks and other thermal metamaterial with novel functions such as concentrators and rotators have been realized in experiments [5].

2.5 Exercises and Solutions

Exercises

1. Prove that Christoffel symbols Γ^v_{vu} satisfy

$$\Gamma^v_{vu} = \frac{1}{\sqrt{g}}\frac{\partial\sqrt{g}}{\partial x^u}. \qquad (2.53)$$

2. Prove transformation theory can be applied to the convection-diffusion equation

$$\frac{\partial\rho c}{\partial t} = \nabla\cdot(\mathbf{D}\nabla c) - \nabla\cdot(\mathbf{s}c) \qquad (2.54)$$

and give the transformation rule for diffusivity tensor $\mathbf{D}$ and velocity $\mathbf{s}$.

3. Calculate the transformation matrix $\dfrac{\mathbf{JJ}^\mathsf{T}}{\det\mathbf{J}}$ for thermal concentrators (which mean guiding heat flux into a central region without disturbing the external temperature signals [5]):

$$r' = \frac{R_1}{R_2}r \text{ as } r < R_2,$$
$$r' = \frac{R_1-R_2}{R_3-R_2}R_3 + \frac{R_3-R_1}{R_3-R_2}r \text{ as } R_2 < r < R_3. \qquad (2.55)$$

Solutions

1. **Proof**: Since

$$g = \left|[g_{uv}]\right|, \qquad (2.56)$$

it is easy to prove

$$\sqrt{g} = \sqrt{|[g_{uv}]|} = (\boldsymbol{g}_u \times \boldsymbol{g}_v) \cdot \boldsymbol{g}_w, \tag{2.57}$$

which means $\sqrt{g}$ is the volume of the parallelepiped whose edges are $\boldsymbol{g}_u, \boldsymbol{g}_v, \boldsymbol{g}_w$. Then, according to Christoffel symbols $\Gamma^w_{vu} = \frac{\partial \boldsymbol{g}_u}{\partial x^v} \cdot \boldsymbol{g}^w$, we obtain (here $i = u, v, w$)

$$
\begin{aligned}
\frac{\partial \sqrt{g}}{\partial x^i} &= \frac{(\boldsymbol{g}_u \times \boldsymbol{g}_v) \cdot \boldsymbol{g}_w}{\partial x^i} \\
&= \left(\frac{\partial \boldsymbol{g}_u}{\partial x^i} \times \boldsymbol{g}_v\right) \cdot \boldsymbol{g}_w + \left(\boldsymbol{g}_u \times \frac{\partial \boldsymbol{g}_v}{\partial x^i}\right) \cdot \boldsymbol{g}_w + (\boldsymbol{g}_u \times \boldsymbol{g}_v) \cdot \frac{\partial \boldsymbol{g}_w}{\partial x^i} \\
&= (\Gamma^k_{ui} \boldsymbol{g}_k \times \boldsymbol{g}_v) \cdot \boldsymbol{g}_w + (\boldsymbol{g}_u \times \Gamma^k_{vi} \boldsymbol{g}_k) \cdot \boldsymbol{g}_w + (\boldsymbol{g}_u \times \boldsymbol{g}_v) \cdot \Gamma^k_{wi} \boldsymbol{g}_k \quad (2.58) \\
&= (\Gamma^u_{ui} \boldsymbol{g}_u \times \boldsymbol{g}_v) \cdot \boldsymbol{g}_w + (\boldsymbol{g}_u \times \Gamma^v_{vi} \boldsymbol{g}_v) \cdot \boldsymbol{g}_w + (\boldsymbol{g}_u \times \boldsymbol{g}_v) \cdot \Gamma^w_{wi} \boldsymbol{g}_w \\
&= \Gamma^j_{ji} \left[(\boldsymbol{g}_u \times \boldsymbol{g}_v) \cdot \boldsymbol{g}_w\right] \\
&= \Gamma^j_{ji} \sqrt{g}.
\end{aligned}
$$

Finally we can complete the proof by taking $j = v$ and $i = u$, and obtain

$$\Gamma^v_{vu} = \frac{1}{\sqrt{g}} \frac{\partial \sqrt{g}}{\partial x^u}. \tag{2.59}$$

2. **Proof**: First we note that the term $\boldsymbol{D}\nabla c$ can be strictly written as $\boldsymbol{D} \cdot \nabla c$ and we have proven that

$$\nabla \cdot [\boldsymbol{D}(c) \cdot \nabla c] = \frac{1}{\sqrt{g}} \partial_u \left[\sqrt{g} D^{uv} \partial_v c\right]. \tag{2.60}$$

Similarly, the term $-\nabla \cdot (s c)$ under a geometric transformation is

$$
\begin{aligned}
-\nabla \cdot (s c) &= -\boldsymbol{g}^w \cdot \frac{\partial}{\partial x^w} (s^u \boldsymbol{g}_u c) \\
&= -\left(\frac{\partial s^u c}{\partial x^u} + \Gamma^w_{wu} s^u c\right) \\
&= -\left[\frac{\partial s^u c}{\partial x^u} + \frac{1}{\sqrt{g}} (\partial_u \sqrt{g}) s^u c\right] \\
&= -\frac{1}{\sqrt{g}} \partial_u (\sqrt{g} s^u c),
\end{aligned}
\tag{2.61}
$$

so the total equation is transformed as

$$\frac{\partial \sqrt{g} \rho c}{\partial t} - \partial_u (\sqrt{g} D^{uv} \partial_v c) + \partial_u (\sqrt{g} s^u c) = 0. \tag{2.62}$$

Still we can obtain the transformed diffusivity matrix $\left[\tilde{\mathbf{D}}\right]$

$$[\tilde{\mathbf{D}}] = \frac{\mathbf{J}[\mathbf{D}]\mathbf{J}^{\mathsf{T}}}{\det \mathbf{J}}. \tag{2.63}$$

In addition, we have

$$s = s^u g_u = s^i i, \tag{2.64}$$

which implies that

$$s^u \mathbf{J}_{iu}^{-1} i = s^i i. \tag{2.65}$$

So the transformed velocity $\tilde{s}$ is

$$\tilde{s} = \frac{\mathbf{J}^{\mathsf{T}} s}{\det \mathbf{J}}. \tag{2.66}$$

3. **Solution**: We note that this can be seen as a geometric transformation which squeezes the region $0 < r < R_2$ to $r' < R_1$ and then stretches the region $R_2 < r < R_3$ to $R_1 < r' < R_3$. The physical field in $r' < R_1$ should be enlarged (concentrated). In $r' < R_1$ the Jacobian matrices are

$$\mathbf{J}_1 = \begin{pmatrix} \frac{R_1}{R_2} & 0 \\ 0 & \frac{R_1}{R_2} \end{pmatrix} \tag{2.67}$$

and

$$\frac{\mathbf{J}_1 \mathbf{J}_1^{\mathsf{T}}}{\det \mathbf{J}_1} = \begin{pmatrix} 1 & 0 \\ 0 & 1 \end{pmatrix}. \tag{2.68}$$

In $R_1 < r' < R_3$, we have

$$\mathbf{J}_2 = \begin{pmatrix} \cos\theta & -r'\sin\theta \\ \sin\theta & r'\cos\theta \end{pmatrix} \begin{pmatrix} \frac{R_3-R_1}{R_3-R_2} & 0 \\ 0 & 1 \end{pmatrix} \begin{pmatrix} \cos\theta & \sin\theta \\ -\frac{\sin\theta}{r} & \frac{\cos\theta}{r} \end{pmatrix}. \tag{2.69}$$

Denoting the transformation matrix in Cartesian coordinate system as

$$\mathbf{T}_2 = \frac{\mathbf{J}_2 \mathbf{J}_2^{\mathsf{T}}}{\det \mathbf{J}_2} = \begin{pmatrix} T_{xx} & T_{xy} \\ T_{yx} & T_{yy} \end{pmatrix}, \tag{2.70}$$

finally we get

$$T_{xx} = \frac{r' + R_3 \frac{R_2-R_1}{R_3-R_2}}{r'} \cos^2\theta + \frac{r'}{r' + R_3 \frac{R_2-R_1}{R_3-R_2}} \sin^2\theta,$$

$$T_{xy} = T_{yx} = \frac{r' + R_3 \frac{R_2-R_1}{R_3-R_2}}{r'} \sin\theta \cos\theta - \frac{r'}{r' + R_3 \frac{R_2-R_1}{R_3-R_2}} \sin\theta \cos\theta, \tag{2.71}$$

$$T_{yy} = \frac{r' + R_3 \frac{R_2-R_1}{R_3-R_2}}{r'} \sin^2\theta + \frac{r'}{r' + R_3 \frac{R_2-R_1}{R_3-R_2}} \cos^2\theta.$$

References

1. Fan, C.Z., Gao, Y., Huang, J.P.: Shaped graded materials with an apparent negative thermal conductivity. Appl. Phys. Lett. **92**, 251907 (2008)
2. Chen, T.Y., Weng, C.N., Chen, J.S.: Cloak for curvilinearly anisotropic media in conduction. Appl. Phys. Lett. **93**, 114103 (2008)
3. Li, Y., Shen, X.Y., Wu, Z.H., Huang, J.Y., Chen, Y.X., Ni, Y.S., Huang, J.P.: Temperature-dependent transformation thermotics: from switchable thermal cloaks to macroscopic thermal diodes. Phys. Rev. Lett. **115**, 195503 (2015)
4. Shen, X.Y., Li, Y., Jiang, C.R., Huang, J.P.: Temperature trapping: energy-free maintenance of constant temperatures as ambient temperature gradients change. Phys. Rev. Lett. **117**, 055501 (2016)
5. Narayana, S., Sato, Y.: Heat flux manipulation with engineered thermal materials. Phys. Rev. Lett. **108**, 214303 (2012)

Chapter 3
Transformation Thermotics for Thermal Conduction and Convection

Abstract In this chapter, we extend the theory of transformation thermotics for thermal conduction described in Chap. 2 to the case of thermal convection. We adopt a model of creeping flow in porous media where Darcy's law is valid and heat flux comes from both thermal conduction and convection. Here the transformation theory is established on a set of equations governing the heat and mass transfer in porous media. We investigate both the steady and transient cases and design a cloak, a concentrator and a rotator for model applications.

Keywords Transformation thermotics · Thermal convection · Porous media · Darcy's law

3.1 Opening Remarks

Besides thermal conduction and thermal radiation, thermal convection is another basic form of heat transport. Thermal convection happens in moving fluids and energy is transferred due to both the movement of mass and the temperature gradient in space. Since we must consider heat and mass transfer together, modulating heat flow in thermal convection could be much more complicated than in thermal conduction. In this chapter, we develop the transformation theory to treat thermal convection.

3.2 Transforming Thermal Convection: Steady Regime

First we consider the steady cases. The governing equation for heat transfer is

$$\rho C_p \nabla \cdot (\boldsymbol{v} T) = \nabla \cdot (\boldsymbol{\kappa} \nabla T), \tag{3.1}$$

which is the convection-diffusion equation. Here ρ, C_p, and $\boldsymbol{\kappa}$ are respectively the density, specific heat at constant pressure, and thermal conductivity tensor of fluid

© Springer Nature Singapore Pte Ltd. 2020
J.-P. Huang, *Theoretical Thermotics*,
https://doi.org/10.1007/978-981-15-2301-4_3

materials. Also $\boldsymbol{v}$ is the velocity of fluids. Compared with conduction, the advection term $\rho C_p \nabla \cdot (\boldsymbol{v} T)$ is added.

In addition, we should consider the governing equations for mass transfer, meaning the movement of fluids. The state of the fluid is determined by Navier-Stokes equation and the equation of continuity

$$(\boldsymbol{v} \cdot \nabla)\boldsymbol{v} = -\frac{1}{\rho}\nabla p + \frac{\beta}{\rho}\nabla \cdot \nabla \boldsymbol{v}, \tag{3.2}$$

$$\nabla \cdot \boldsymbol{v} = 0. \tag{3.3}$$

Here β denotes the dynamic viscosity and p denotes the pressure. For simplicity, here we assume that the flow is laminar, Newtonian and the density doesn't defend on temperature.

Now we must deal with a set of coupled equations. The transformation theory can apply to more than one equation at the same time, such as Maxwell equations, if all the equations can keep form invariant under coordinate transformation. Equation (3.1) or the the convection-diffusion equation is proved satisfying such requirement (Exercise 2 of Chap. 2 is for transient equation and the conclusion can be directly applied to steady-state equation), and Eq. (3.1) is transformed as

$$- \partial_u(\sqrt{g}\eta^{uv}\partial_v T) + \rho C_p \partial_u(\sqrt{g}v^u T) = 0. \tag{3.4}$$

Still we can obtain the transformed thermal conductivity matrix $\left[\kappa'\right]$

$$\left[\kappa'\right] = \frac{\mathbf{J}[\kappa]\mathbf{J}^\mathsf{T}}{\det \mathbf{J}}. \tag{3.5}$$

In addition, we have

$$\boldsymbol{v} = v^u \boldsymbol{g}_u = v^i \boldsymbol{i} \tag{3.6}$$

which implies that

$$v^u \mathbf{J}_{iu}^{-1}\boldsymbol{i} = v^i \boldsymbol{i} \tag{3.7}$$

so the transformed velocity is

$$\boldsymbol{v}' = \frac{\mathbf{J}\boldsymbol{v}}{\det \mathbf{J}}. \tag{3.8}$$

Another proof can be found in Refs. [1, 2], which resort to some techniques of functional analysis.

Then, we can easily find that the equation of continuity can also be transformed as

$$\partial_u(\sqrt{g}v^u) = 0, \tag{3.9}$$

which is consistent with Eq. (3.8).

Unfortunately, the Navier-Stokes equation does not meet the requirement of form-invariance [3]. Researchers [3] replace it with Darcy's law in porous media, and Darcy's law is valid when the Reynolds number is small or the flow is creeping. Here Darcy's law reads

$$\nabla p + \frac{\beta}{k} \boldsymbol{v} = 0, \tag{3.10}$$

where k is the permeability. Taking Eq. (3.3) into Eq. (3.10) we have

$$\nabla \cdot (\frac{k}{\beta} \nabla p) = 0, \tag{3.11}$$

which has the same form as Fourier's law, which are both diffusion equation, so the form-invariance under coordinate transformation is obvious. We can directly write the transformation rule for $\lambda = \frac{k}{\beta}$ as

$$[\lambda'] = \frac{\mathbf{J}[\lambda]\mathbf{J}^\mathsf{T}}{\det \mathbf{J}}. \tag{3.12}$$

Without causing confusion, we neglect the symbol $[]$ for matrix in the following. Since a porous medium is a mixture of solid and fluid, Eq. (3.1) can be rewritten as

$$\rho_f C_{p,f}(\boldsymbol{v} \cdot \nabla T) = \nabla \cdot (\kappa_m \nabla T). \tag{3.13}$$

κ_m is the effective thermal conductivity of porous media and can be written as

$$\kappa_m = (1 - \phi)\kappa_s + \phi\kappa_f, \tag{3.14}$$

by the volume-averaging method. Here ϕ denotes the porosity and κ_f and κ_s are the thermal conductivity of fluid and solid respectively. Since it's difficult to directly engineer the properties of fluid, we only transform the solid materials and finally we have the following rules (in the following part we don't distinguish rank-2 tensor and matrix in Cartesian coordinate system) [4]:

$$\begin{cases} \kappa'_m = \dfrac{\mathbf{J}\kappa_m\mathbf{J}^\mathsf{T}}{\det \mathbf{J}} \\ \kappa'_f = \kappa_f \\ \kappa'_s = \dfrac{\kappa'_m - \phi\kappa_f}{1 - \phi}. \end{cases} \tag{3.15}$$

We should point out that the transformation of velocity is automatically reached if we transform λ.

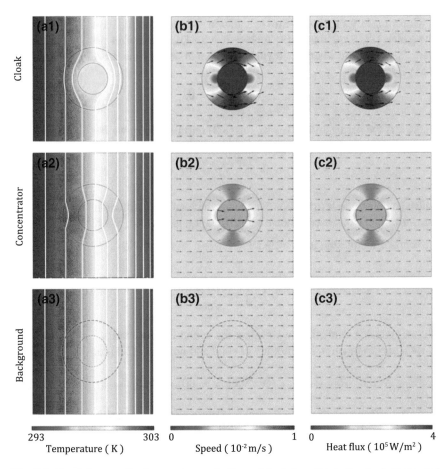

Fig. 3.1 Simulation results (temperature, velocity and heat flux distributions) of cloaks and concentrators for steady thermal convection. Adapted from Ref. [4]

In Fig. 3.1, we show the simulation results for some basic applications of steady transforming thermal convection [4]. This figure illustrates the temperature, velocity and heat flux distributions of thermal cloak and thermal concentrator. In (a1–a3), we plot the isotherms by using white lines. In (b1–b3), the color represents the speed and the black arrows show the direction of velocity. Similarly, in (c1–c3), the color denotes the volume of heat flux and the black arrows show the direction. We can see, for example, a cloak for thermal convection actually makes it impossible to judge if there is an object inside the cloak from detecting the external thermal or velocity filed. Essentially, it is a bi-functional cloak with the functions of thermal cloaking and hydrodynamic cloaking.

3.3 Transforming Thermal Convection: Transient Regime

Now we consider the cases when the temperature, velocity and density all vary with time. Firstly, the equation of continuity Eq. (3.3) can be rewritten as

$$\frac{\partial(\phi\rho_f)}{\partial t} + \nabla \cdot (\rho_f \vec{v}) = 0, \tag{3.16}$$

where ρ_f denotes the density of the fluid material. Then Eq. (3.13) is given by

$$\frac{\partial(\rho_m C_{p,m})T}{\partial t} + \nabla \cdot \left(\rho_f C_{p,f} \boldsymbol{v} T\right) = \nabla \cdot \left(\kappa_m \nabla T\right), \tag{3.17}$$

where

$$\rho_m C_{p,m} = (1 - \phi)\left(\rho_s C_{p,s}\right) + \phi\left(\rho_f C_{p,f}\right) \tag{3.18}$$

and $\rho_m C_{p,m}$ is the product of the density and specific heat of the porous media while $\rho_s C_{p,s}$ and $\rho_f C_{p,f}$ denote the counterpart of solid and fluid, respectively. Combining Eqs. (3.16) and (3.17) we have

$$(\rho_m C_{p,m})\frac{\partial T}{\partial t} + \rho_f C_{p,f}(\boldsymbol{v} \cdot \nabla T) = \nabla \cdot \left(\kappa_m \nabla T\right). \tag{3.19}$$

It is easy to find the transformation rules for conductivity and λ [5]

$$\begin{cases} \kappa'_m = \dfrac{\mathbf{J}\kappa_m \mathbf{J}^{\mathsf{T}}}{\det \mathbf{J}} \\ \kappa'_f = \kappa_f \\ \kappa'_s = \dfrac{\kappa'_m - \phi'\kappa_f}{1 - \phi'} \\ \lambda'_m = \dfrac{\mathbf{J}\lambda_m \mathbf{J}^{\mathsf{T}}}{\det \mathbf{J}} \end{cases}. \tag{3.20}$$

Here ϕ is also transformed indicated from Eq. (3.16) since ρ_f is kept unchanged, and with Eq. (3.19) we have [5]

$$\begin{cases} \phi' = \frac{\phi}{\det \mathbf{J}} \\ \left(\rho_m C_{p,m}\right)' = \frac{1}{\det \mathbf{J}}\rho_m C_{p,m} \end{cases}. \tag{3.21}$$

Finally we have [5]

$$\begin{cases} \left(\rho_f C_{p,f}\right)' = \rho_f C_{p,f} \\ \left(\rho_s C_{p,s}\right)' = \frac{1-\phi}{\det \mathbf{J} - \phi}\rho_s C_{p,s} \end{cases}. \tag{3.22}$$

Finally we talk about the transient version of Darcy's law. In most cases, the relaxation process is very fast in porous media so we can still use Eq. (3.10). Also, the afore-mentioned discussion indicates the validity of Eq. (3.8). The main difference between steady and transient regimes is that both the porosity and the product of density and specific heat of solid should be transformed in transient cases, which makes modulating thermal convection more complicated. In addition, for both steady and transient regimes, we focus on forced convection. When considering natural or mixed convection, we must consider the effect of gravity and it should be difficult to apply the transformation theory strictly because gravity acceleration is hard to be manipulated [5].

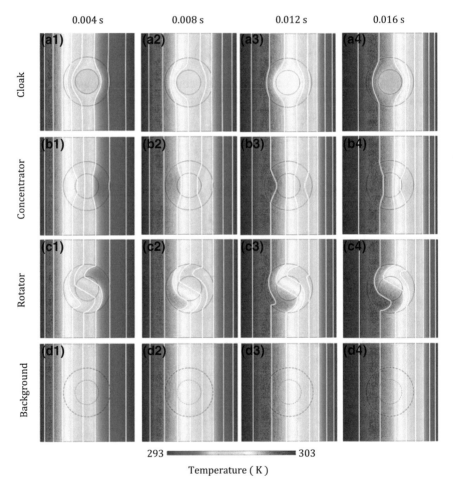

Fig. 3.2 Simulation results (temperature distributions) of cloaks, concentrators and rotators for transient thermal convection. Adapted from Ref. [5]

In Figs. 3.2, 3.3 and 3.4, we also show the simulation results for some applications of transient transformation convection [5]. We design a thermal cloak, concentrator and rotator. Figure 3.2 shows the temperature distributions varying with time while Figs. 3.3 and 3.4 show the velocity and heat flux distributions, respectively. Again we can see the devices can work as thermal and hydrodynamic cloaks, concentrators, or rotators.

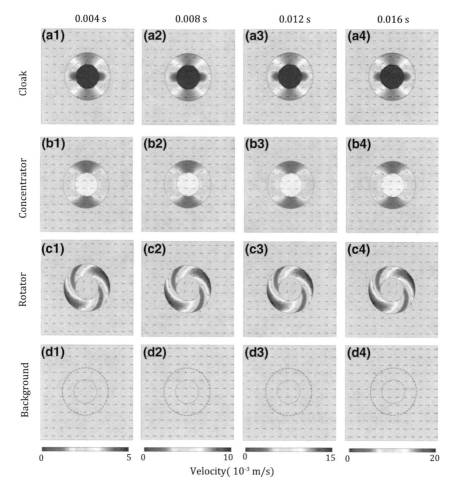

Fig. 3.3 Simulation results (velocity distribution) of cloaks, concentrators and rotators for transient thermal convection. Adapted from Ref. [5]

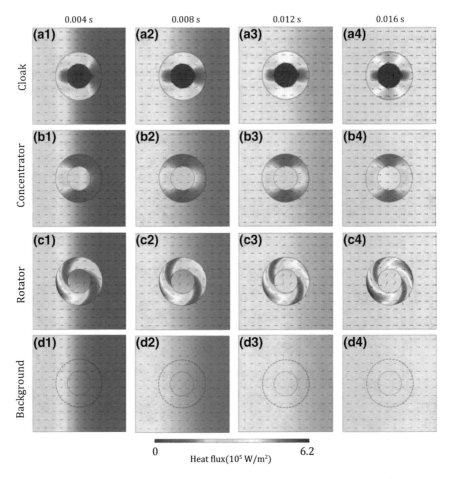

Fig. 3.4 Simulation results (heat flux distribution) of cloaks, concentrators and rotators for transient thermal convection. Adapted from Ref. [5]

3.4 Exercises and Solutions

Exercises

1. Calculate the transformation matrix for thermal rotators:

$$\begin{cases} \theta' = \theta + \theta_0, \ r < R_1 \\ \theta' = ar + b, \ R_1 < r < R_2 \end{cases}, \tag{3.23}$$

where $a = \frac{\theta_0}{R_1 - R_2}$, $b = \theta + \frac{R_2}{R_2 - R_1}\theta_0$ and $R_1 < r < R_2$.

2. What is the form of Darcy's law when considering gravity in natural convection?

Solutions

1. **Solution**: For region $r < R_1$, the Jacobian matrix is expressed as

$$\mathbf{J}_1 = \begin{pmatrix} \cos\theta_0 & -\sin\theta_0 \\ \sin\theta_0 & \cos\theta_0 \end{pmatrix} \tag{3.24}$$

by which we can expect the velocity and heat flux to be rotated by an angle of θ_0. We have an identity transformation matrix as

$$\frac{\mathbf{J}_1\mathbf{J}_1^\mathsf{T}}{\det \mathbf{J}_1} = \begin{pmatrix} 1 & 0 \\ 0 & 1 \end{pmatrix}, \tag{3.25}$$

so in this region we don't need to do any transformation. For region $R_1 < r < R_2$, we can also obtain a unitary Jacobian matrix as

$$\mathbf{J}_2 = \begin{pmatrix} \cos\theta' & -r\sin\theta' \\ \sin\theta' & r\cos\theta' \end{pmatrix} \begin{pmatrix} 1 & 0 \\ a & 1 \end{pmatrix} \begin{pmatrix} \cos\theta & \sin\theta \\ -\frac{\sin\theta}{r} & \frac{\cos\theta}{r} \end{pmatrix}. \tag{3.26}$$

Then the transformation matrix in Cartesian coordinate system is

$$\frac{\mathbf{J}_2\mathbf{J}_2^\mathsf{T}}{\det \mathbf{J}_2} = \begin{pmatrix} \cos\theta' & -r\sin\theta' \\ \sin\theta' & r\cos\theta' \end{pmatrix} \begin{pmatrix} 1 & a \\ a & a^2 + \frac{1}{r^2} \end{pmatrix} \begin{pmatrix} \cos\theta' & -r\sin\theta' \\ \sin\theta' & r\cos\theta' \end{pmatrix}^\mathsf{T}. \tag{3.27}$$

Again we denote the transformation matrix as

$$\mathbf{T}_2 = \frac{\mathbf{J}_2\mathbf{J}_2^\mathsf{T}}{\det \mathbf{J}_2} = \begin{pmatrix} T_{xx} & T_{xy} \\ T_{yx} & T_{yy} \end{pmatrix}. \tag{3.28}$$

Finally we obtain

$$T_{xx} = \cos^2\theta' - ar\sin(2\theta') + (1 + a^2r^2)\sin^2\theta',$$

$$T_{xy} = T_{yx} = ar\cos(2\theta') - \frac{1}{2}a^2r^2\sin(2\theta'), \tag{3.29}$$

$$T_{yy} = \sin^2\theta' + ar\sin(2\theta') + (1 + a^2r^2)\cos^2\theta'.$$

2. **Solution**: Denoting gravitational acceleration as g, we may achieve

$$v = -\frac{\beta}{k}\left(\nabla p - \rho_f g\right). \tag{3.30}$$

References

1. Guenneau, S., Puvirajesinghe, T.M.: Fick's second law transformed: one path to cloaking in mass diffusion. J. Roy. Soc. Interface **10**, 20130106 (2013)
2. Guenneau, S., Petiteau, D., Zerrad, M., Amra, C., Puvirajesinghe, T.: Transformed Fourier and Fick equations for the control of heat and mass diffusion. AIP Adv. **5**, 053404 (2015)
3. Urzhumov, Y.A., Smith, D.R.: Fluid flow control with transformation media. Phys. Rev. Lett. **107**, 074501 (2011)
4. Dai, G.L., Shang, J., Huang, J.P.: Theory of transformation thermal convection for creeping flow in porous media: cloaking, concentrating, and camouflage. Phys. Rev. E **97**, 022129 (2018)
5. Dai, G.L., Huang, J.P.: A transient regime for transforming thermal convection: cloaking, concentrating, and rotating creeping flow and heat flux. J. Appl. Phys. **124**, 235103 (2018)

Chapter 4
Transformation Thermotics for Thermal Conduction and Radiation

Abstract Apart from conduction and convection, thermal radiation is the third fundamental mechanism of heat transfer. Any object with a non-zero temperature can emit thermal radiation through electromagnetic waves, which means the thermal energy is converted into electromagnetic energy. From the insight of electromagnetic waves, the transport of thermal radiation can be naturally incorporated into the framework of transformation optics. However, one lacks a model dealing with heat flux resulting from thermal conduction and radiation at the same time and also it is necessary to efficiently control the thermal radiation described by the Stefan-Boltzmann law by using the transformation theory. For this purpose, in this chapter we introduce a radiation model called Rosseland diffusion approximation [1] and establish a transformation theory for thermal radiation associated with conduction within a single framework [2]. As model applications, we also use the theory to develop thermal cloaking, concentrating, and rotating for thermal radiation.

Keywords Transformation thermotics · Thermal radiation · Rosseland diffusion approximation · Cloak · Concentrator · Rotator

4.1 Rosseland Diffusion Approximation

Here we give a brief derivation of Rosseland diffusion approximation, following the framework in Ref. [3]. Consider a beam of electromagnetic waves (light) travels through a medium with a mass density ρ. Generally, the intensity of light should be reduced as the media is not absolutely transparent (due to absorption and scattering). For a monochromatic light with a frequency ν, we assume it travels along the z direction. Now we introduce the concept of Rosseland diffusion approximation, which is valid in optically thick media. This approximation indicates that the mean free path of photons is far smaller than the thickness of media. A typical example is the interiors of stars where the mean free path of photons is much smaller than the characteristic lengths of temperature and particle density gradients. We can write a differential equation for the local intensity of light I_ν as

$$\cos\theta \frac{dI_\nu}{dz} = \beta_\nu(S_\nu - I_\nu) \tag{4.1}$$

© Springer Nature Singapore Pte Ltd. 2020
J.-P. Huang, *Theoretical Thermotics*,
https://doi.org/10.1007/978-981-15-2301-4_4

where S_ν is the source intensity and β_ν is a coefficient measuring attenuation. Usually we can write it as $\beta_\nu = (\rho\alpha_\nu)^{-1}$ and α_ν is called the Rosseland opacity. We can see $I_\nu(z)$ demonstrates a form of e-exponential decaying as $I_\nu(z) = I_\nu(z=0)e^{-\beta_\nu z}$ if the source function equals zero and we can see the opacity means $100\% \left(1 - \frac{I_\nu(z)}{I_\nu(z=0)}\right)$. In the Rosseland diffusion approximation, we assume that S_ν and $I_\nu(z)$ are both similar to black-body radiation function $B_\nu(T)$ (T denotes temperature):

$$I_\nu^{(0)} \approx S_\nu \approx B_\nu(T), \tag{4.2}$$

where $I_\nu^{(0)}$ is the zero-order solution of local intensity and

$$B_\nu(\nu, T) = \frac{2h\nu^3}{c^2} \frac{1}{e^{\frac{h\nu}{k_B T}} - 1}. \tag{4.3}$$

Here c is the speed of light in the medium, h is the Planck constant, k_B is the Boltzmann constant and it's easy to check that $B_\nu(T)$ satisfies

$$\int_0^\infty B_\nu(T)d\nu = n^2 \frac{\sigma}{\pi} T^4. \tag{4.4}$$

Here σ is the Stefan-Boltzmann constant $\left(5.67 \times 10^{-8}\,\mathrm{Wm^{-2}K^{-4}}\right)$ and n is the relative refraction index. Take $I_\nu \approx I_\nu^{(0)} + I_\nu^{(1)}$ and we will find the first-order solution of local intensity writes

$$\begin{aligned} I_\nu^{(1)} &= B_\nu(T) - \cos\theta\beta_\nu^{-1}\frac{\partial B_\nu(T)}{\partial z} \\ &= B_\nu(T) - \cos\theta\beta_\nu^{-1}\frac{\partial B_\nu(T)}{\partial T}\frac{\partial T}{\partial z}. \end{aligned} \tag{4.5}$$

Then we turn to calculate the radiative heat flux (density) $\boldsymbol{J}_{rad}$:

$$\boldsymbol{J}_{rad} = \int_0^\infty d\nu \int_{\Omega_0} d\Omega\, I_\nu(T)\cos\theta, \tag{4.6}$$

where Ω_0 is the unit hemispherical surface. Notice that $B_\nu(T)$ is isotropic and make a variable replacement $\mu = \cos\theta$, we have

$$\begin{aligned} \int_{\Omega_0} d\Omega\, I_\nu(T)\cos\theta &= \int_{\Omega_0} \cos\theta\left[2B_\nu(T) - \cos\theta\beta_\nu^{-1}\frac{\partial B_\nu(T)}{\partial T}\frac{\partial T}{\partial z}\right]d\Omega \\ &= \int_{\Omega_0} \cos^2\theta\left[-\beta_\nu^{-1}\frac{\partial B_\nu(T)}{\partial T}\frac{\partial T}{\partial z}\right]\sin\theta d\theta d\psi \\ &= -2\pi\beta_\nu^{-1}\frac{\partial B_\nu(T)}{\partial T}\frac{\partial T}{\partial z}\int_{-1}^1 \mu^2 d\mu \\ &= -\frac{4\pi}{3}\beta_\nu^{-1}\frac{\partial B_\nu(T)}{\partial T}\frac{\partial T}{\partial z}. \end{aligned} \tag{4.7}$$

Then we can see over the whole range of frequency,

$$\boldsymbol{J}_{rad} = -\frac{4\pi}{3}\frac{\partial T}{\partial z}\int_0^\infty \frac{\partial B_\nu(T)}{\partial T}\frac{1}{\beta_\nu}d\nu, \tag{4.8}$$

and we can define an average absorption coefficient or so-called Rosseland mean attenuation coefficient β as

$$\beta^{-1}\int_0^\infty \frac{\partial B_\nu(T)}{\partial T}d\nu = \int_0^\infty \frac{\partial B_\nu(T)}{\partial T}\frac{1}{\beta_\nu}d\nu. \tag{4.9}$$

Note Eq. (4.4), we can write the Rosseland mean attenuation coefficient as

$$\beta^{-1} = \frac{\pi\int_0^\infty \frac{\partial B_\nu(T)}{\partial T}\frac{1}{\beta_\nu}d\nu}{4n^2\sigma T^3}. \tag{4.10}$$

Finally, the radiative flux $\boldsymbol{J}_{rad}$ according to the Rosseland diffusion approximation is

$$\boldsymbol{J}_{rad} = -\frac{16}{3}\beta^{-1}n^2\sigma T^3 \cdot \nabla T. \tag{4.11}$$

4.2 Transforming Thermal Radiation

Here we consider a transient thermal transport process with both thermal radiation and conduction; see Fig. 4.1a. When the thermal transport process is passive, the dominant equation becomes

$$\rho C\frac{\partial T}{\partial t} + \nabla \cdot (\boldsymbol{J}_{rad} + \boldsymbol{J}_{con}) = 0, \tag{4.12}$$

where ρ and C are the density and heat capacity of the participating media respectively. The conductive flux $\boldsymbol{J}_{con}$ is determined by the Fourier law ($-\kappa \cdot \nabla T$), where κ is thermal conductivity. To check the form-invariant under a coordinate transformation of Eq. (4.12), we denote

$$\tau = -\frac{16}{3}\beta^{-1}n^2\sigma \tag{4.13}$$

and thus the radiation term in Eq. (4.12) is

$$\nabla \cdot \boldsymbol{J}_{rad} = \nabla \cdot (\tau T^3 \cdot \nabla T). \tag{4.14}$$

The other terms in Eq. (4.12) are just the same as the transient Fourier's law in heat conduction and again we have the transformation rules for κ and ρC as

$$[\tilde{\kappa}(T)] = \frac{\mathbf{J}[\kappa(T)]\mathbf{J}^\mathsf{T}}{\det \mathbf{J}} \tag{4.15}$$

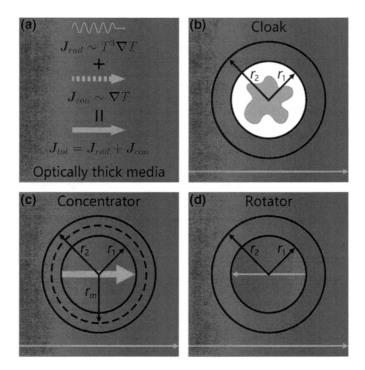

Fig. 4.1 Schematic graph showing **a** thermal transport process, **b** thermal cloak, **c** thermal concentrator, and **d** thermal rotator. In **a**, the wavy arrow, dashed arrow, and solid arrow denote radiative flux, conductive flux, and total flux, respectively. Adapted from Ref. [2]

and

$$\tilde{\rho}\tilde{C} = \frac{\rho C}{\det \mathbf{J}} \tag{4.16}$$

where $\mathbf{J}$ is the Jacobian transformation matrix from the Cartesian coordinate system to the curvilinear coordinate system. So, we only need to check the radiation term under coordinate transformation. In any curvilinear coordinate system with a contravariant basis $\{g^u, g^v, g^w\}$, covariant basis $\{g_u, g_v, g_w\}$ and corresponding contravariant components $\{x^u, x^v, x^w\}$, the radiation term can be written as

$$
\begin{aligned}
\nabla \cdot \left[\tau T^3 \cdot \nabla T \right] &= g^w \cdot \frac{\partial}{\partial x^w} \left[\tau^{uv} T^3 g_u \otimes g_v \cdot g^l \frac{\partial T}{\partial x^l} \right] \\
&= g^w \cdot \frac{\partial}{\partial x^w} \left[\tau^{uv} T^3 g_u \frac{\partial T}{\partial x^v} \right] \\
&= \frac{\partial \tau^{uv} T^3}{\partial x^u} \frac{\partial T}{\partial x^v} + \frac{\partial^2 T}{\partial x^u \partial x^v} \tau^{uv} T^3 + g^w \cdot \frac{\partial g_u}{\partial x^w} \left[\tau^{uv} T^3 \frac{\partial T}{\partial x^v} \right] \\
&= \partial_u \left[\tau^{uv} T^3 \partial_v T \right] + \Gamma^w_{wu} \tau^{uv} T^3 \partial_v T
\end{aligned}
$$

$$= \partial_u \left[\tau^{uv} T^3 \partial_v T \right] + \frac{1}{\sqrt{g}} (\partial_u \sqrt{g}) \tau^{uv} T^3 \partial_v T$$

$$= \frac{1}{\sqrt{g}} \partial_u \left[\sqrt{g} \tau^{uv} T^3 \partial_v T \right], \tag{4.17}$$

where g is the determinant of matrix $g_{ij} = g_u \cdot g_v$. Γ^w_{wu} is the Christoffel symbol represented by $\Gamma^w_{vu} = \frac{\partial g_u}{\partial x^v} \cdot g^w$ and we should remember that $\Gamma^w_{wu} = \frac{1}{\sqrt{g}} \frac{\partial \sqrt{g}}{\partial x^u}$. We can find the radiation term is form-invariant under coordinate transformation if we follow the transformation rule below

$$\left[\tilde{\tau}(T) \right] = \frac{\mathbf{J}\left[\tau(T) \right] \mathbf{J}^\mathsf{T}}{\det \mathbf{J}}. \tag{4.18}$$

In fact, we may see τT^3 as $\eta(T)$ and directly write the transformed matrix $\left[\tilde{\eta}(T) \right] = \frac{\mathbf{J}\left[\eta(T) \right] \mathbf{J}^\mathsf{T}}{\det \mathbf{J}}$ just as the transformation rule of thermal conductivity. In other words, the Rosseland diffusion approximation takes the radiation flux in the form of heat conduction (diffusion) with an effective temperature-dependent thermal conductivity $\eta(T)$. Considering that natural materials have only a small range of relative refraction indexes, we don't transform $\mathbf{n}$. Also, we can't transform the Stefan-Boltzmann constant. Finally we have the transformation rule for the Rosseland mean attenuation coefficient as

$$\left[\tilde{\beta}(T) \right] = \frac{\mathbf{J}\left[\beta(T) \right] \mathbf{J}^\mathsf{T}}{\det \mathbf{J}}. \tag{4.19}$$

Thus we have established a transformation theory for thermal radiation (accompanying with conduction) under the Rosseland diffusion approximation.

Now we can design thermal metamaterial in environments of both radiation and conduction. Figure 4.1b–d illustrates the schematic diagrams for cloaks, concentrators and rotators respectively. We still use the familiar coordinate transformations for cloaks

$$\begin{cases} r' = (r_2 - r_1) r / r_2 + r_1 \\ \theta' = \theta \end{cases}, \tag{4.20}$$

concentrators

$$\begin{cases} r' = r_1 r / r_m \text{ for } r < r_m \\ r' = \left[(r_2 - r_1) r + (r_1 - r_m) r_2 \right] / (r_2 - r_m) \text{ for } r_m < r < r_2 \\ \theta' = \theta \end{cases}, \tag{4.21}$$

and rotators

$$\begin{cases} r' = r \\ \theta' = \theta + \theta_0 \text{ for } r < r_1 \\ \theta' = \theta + \theta_0 (r - r_2) / (r_1 - r_2) \text{ for } r_1 < r < r_2 \end{cases}, \tag{4.22}$$

where θ_0 is rotation degree. It's easy to see that when temperature is high, the radiative flux could raise faster than conductive flux. We choose three different temperature intervals, meaning (300, 320 K), (300, 1000 K) and (300, 4000 K) and set the boundary values as the temperature of low/high source in numerical simulations respectively. Figure 4.2 shows the temperature distributions varying with time for thermal cloaks. The first column (a)–(d) has a cold source of 300 K and a heat source of 320 K so the radiative flux is much smaller than the conductive flux. The second column (e)–(h) has a cold source of 300 K and a heat source of 1000 K so the radiative flux is comparable with the conductive flux. The last column (i)–(l) has a cold source of 300 K and a heat source up to 4000 K so the radiative flux is much larger the conductive flux. Similarly, Figs. 4.3 and 4.4 show the temperature distribution for concentrators and rotators, respectively. We can find for all the three cases, the effects of cloaking, concentrating, and rotating are achieved indeed.

4.3   Exercises and Solutions

Exercises

1. Prove Eq. (4.4), meaning

$$\int_0^\infty B_\nu(T)d\nu = n^2\frac{\sigma}{\pi}T^4. \tag{4.23}$$

Solutions

1. **Solution**: We can rewrite Eq. (4.3)

$$B_\nu(\nu, T) = \frac{2h\nu^3}{c_{vacuum}^2}\frac{n^2}{e^{\frac{h\nu}{k_B T}} - 1}. \tag{4.24}$$

Then

$$\int_0^\infty B_\nu(T)d\nu = \frac{2hn^2}{c_{vacuum}^2}\int_0^\infty \frac{\nu^3}{e^{\frac{h\nu}{k_B T}} - 1}$$
$$= \frac{2hn^2}{c_{vacuum}^2}\left(\frac{k_B T}{h}\right)^4\int_0^\infty \frac{u^3}{e^u - 1}du, \tag{4.25}$$

where the Bose–Einstein integral or the Riemann zeta function $6\zeta(4)$ equals

$$\int_0^\infty \frac{u^3}{e^u - 1}du = \frac{\pi^4}{15}. \tag{4.26}$$

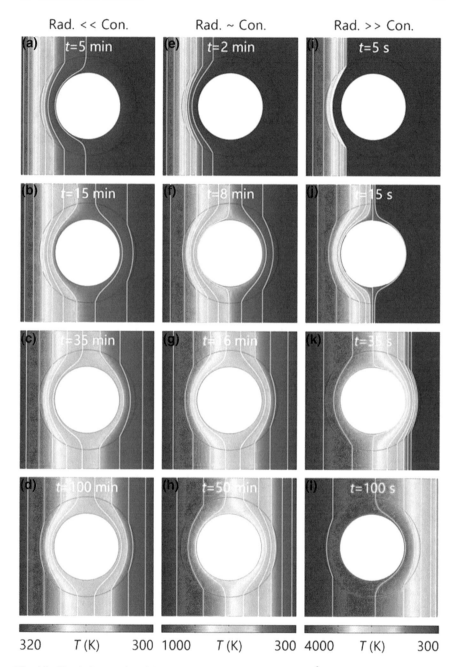

Fig. 4.2 Simulation results of thermal cloak. The size is 10×10 cm^2, $r_1 = 2.4$, $r_2 = 3.6$ cm, and the background parameters are $n_0 = 1$, $\beta_0 = 100$ m^{-1}, $\kappa_0 = 1$ Wm^{-1}K^{-1}, $\rho_0 C_0 = 10^6$ Jm^{-3}K^{-1}. The evolutions over time are demonstrated in **a–d**, **e–h**, and **i–l**, with three different temperature intervals. The cloak parameters are set as required by Eqs. (4.15), (4.16) and (4.19), and the corresponding Jacobian matrix is determined by Eq. (4.20). Adapted from Ref. [2]

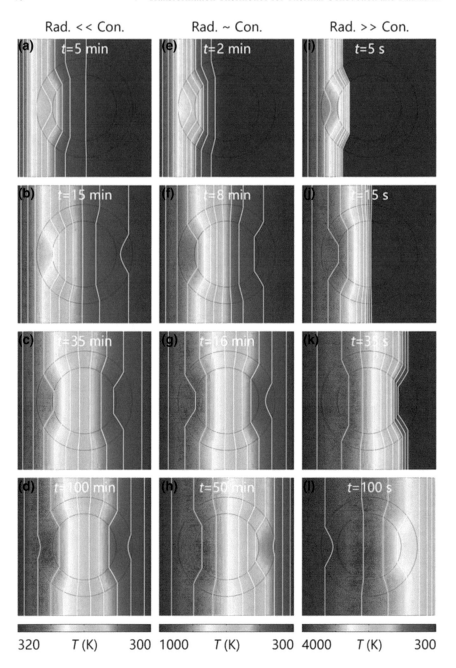

Fig. 4.3 Simulation results of thermal concentrator. The parameters are the same as those for Fig. 4.2, except for the Jacobian transformation matrix (determined by Eq. (4.21) where $r_m = 3.2$ cm). Adapted from Ref. [2]

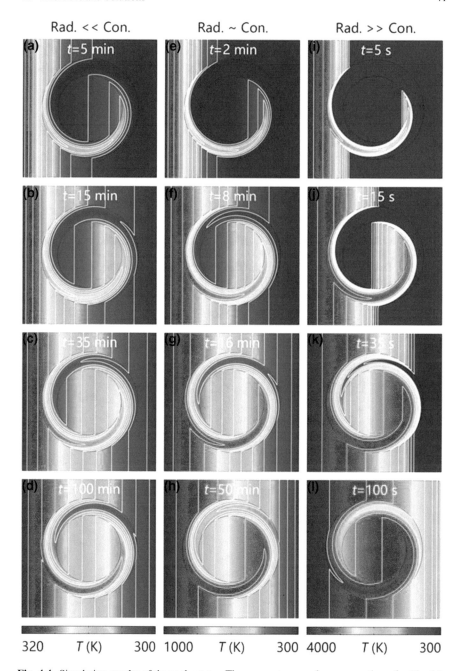

Fig. 4.4 Simulation results of thermal rotator. The parameters are the same as those for Fig. 4.2, except for the Jacobian transformation matrix (determined by Eq. (4.22)). Adapted from Ref. [2]

Define the Stefan–Boltzmann constant as

$$\sigma = \frac{2\pi^5 k_{\mathrm{B}}^4}{15 c_{vacuum}^2 h^3} = 5.670373 \times 10^{-8}\,\mathrm{W\,m^{-2}K^{-4}}, \qquad (4.27)$$

we have

$$\int_0^\infty B_\nu(T)d\nu = \boldsymbol{n}^2 \frac{\sigma}{\pi} T^4. \qquad (4.28)$$

References

1. Rosseland, S.: Theoretical Astrophysics. Oxford University Press, Clarendon, London and New York (1936)
2. Xu, L.J., Dai, G.L., Huang, J.P.: Transformation multithermotics: controlling radiation and conduction simultaneously (submitted) (2019)
3. Howell, J.R., Menguc, M.P., Siegel, R.: Thermal Radiation Heat Transfer, 5th edn. CRC Press, Boca Raton, London, New York (2010)

Chapter 5
Transformation Thermotics for Thermal Conduction, Convection and Radiation

Abstract In Chaps. 2, 3 and 4, we have investigated the transformation of thermal conduction, convection, and radiation. In fact, in our model of transforming thermal convection or radiation, thermal conduction has also been considered; see Chaps. 3 and 4. However, we have not considered the most general case of heat transfer consisting of conduction, convection and radiation together. Here we integrate the relevant content of the previous chapters and give a comprehensive framework of transforming heat transfer (conduction, convection and radiation). In this case, a thermal cloak is also designed as an application.

Keywords Transformation thermotics · Thermal radiation · Thermal convection · Thermal conduction · Cloak

5.1 Transformation Theory

Using the Rosseland diffusion approximation [1], the transient process of heat transfer in pure fluids with thermal conduction, convection, and radiation is governed by

$$\partial(\rho_f C_f T)/\partial t + \nabla \cdot \left(-\kappa_f \cdot \nabla T + \rho_f C_f \boldsymbol{v} T - 16/3\beta_f^{-1} \boldsymbol{n}_f^2 \sigma T^3 \cdot \nabla T\right) = 0,$$

(5.1)

where ρ_f, C_f, κ_f, $\boldsymbol{v}$, β_f, and $\boldsymbol{n}_f$ are density, heat capacity, thermal conductivity, velocity, the Rosseland mean attenuation coefficient, and relative refraction index of the fluid, respectively. σ is the Stefan-Boltzmann constant $\left(=5.67 \times 10^{-8}\ \mathrm{Wm}^{-2}\mathrm{K}^{-4}\right)$. T denotes temperature, and t represents time. Among them, the conductive flux is determined by Fourier's law $\boldsymbol{J}_F = -\kappa_f \cdot \nabla T$; the convective flux is given by $\boldsymbol{J}_C = \rho_f C_f \boldsymbol{v}(T - T_{Ref})$ and T_{Ref} is the reference temperature; the radiative flux given by the Rosseland diffusion approximation is $\boldsymbol{J}_R = -16/3\beta_f^{-1} \boldsymbol{n}_f^2 \sigma T^3 \cdot \nabla T$; and the total flux is the summation of conductive, convective, and radiative flux $\boldsymbol{J}_T = \boldsymbol{J}_F + \boldsymbol{J}_C + \boldsymbol{J}_R$.

Equation (5.1) can keep the form-invariance under a coordinate transformation from the curvilinear space X to the physical space X', and its corresponding Jacobian transformation matrix is denoted as $\mathbf{J}$. Following the proofs in previous chapters, we choose a curvilinear space X with a contravariant basis $\left(\boldsymbol{g}^i, \boldsymbol{g}^j, \boldsymbol{g}^k\right)$ and corresponding contravariant components $\{x^i,\ x^j,\ x^k\}$, and Eq. (5.1) can be rewritten as

$$\sqrt{g}\partial_t(\rho_f C_f T) + \partial_i\left(\sqrt{g}\left(-\kappa_f^{ij}\partial_j T + \rho_f C_f v^i T - \alpha_f^{ij} T^3 \partial_j T\right)\right) = 0, \quad (5.2)$$

where $\alpha_f\ \left(=16/3\beta_f^{-1} n_f^2 \sigma\right)$ is the radiative coefficient, and g is the determinant of the matrix with components $g_{mn} = \mathbf{g}_i \cdot \mathbf{g}_j$, where $(\mathbf{g}_i,\ \mathbf{g}_j,\ \mathbf{g}_k)$ is a covariant basis. Equation (5.2) is expressed in the curvilinear space, which should then be rewritten in the physical space with Cartesian coordinate system $\{x^{i'},\ x^{j'},\ x^{k'}\}$,

$$\sqrt{g}\partial_t(\rho_f C_f T) + \partial_{i'}\left(\sqrt{g}\left(-\kappa_f^{ij}\frac{\partial x^{j'}}{\partial x^j}\partial_{j'} T + \rho_f C_f v^i T - \alpha_f^{ij} T^3 \frac{\partial x^{j'}}{\partial x^j}\partial_{j'} T\right)\right) = 0,$$
$$(5.3)$$

where $\partial x^{i'}/\partial x^i$ and $\partial x^{j'}/\partial x^j$ are just the components of the Jacobian transformation matrix $\mathbf{J}$, and $\sqrt{g} = 1/\det \mathbf{J}$. Again, we can rewrite Eq. (5.3) as

$$\frac{1}{\det \mathbf{J}}\partial_t(\rho_f C_f T) + \partial_{i'}\left(-\frac{\frac{\partial x^{i'}}{\partial x^i}\kappa_f^{ij}\frac{\partial x^{j'}}{\partial x^j}}{\det \mathbf{J}}\partial_{j'} T + \frac{\rho_f}{\det \mathbf{J}}C_f \frac{\partial x^{i'}}{\partial x^i} v^i T - \frac{\frac{\partial x^{i'}}{\partial x^i}\alpha_f^{ij}\frac{\partial x^{j'}}{\partial x^j}}{\det \mathbf{J}}T^3 \partial_{j'} T\right) = 0$$
$$(5.4)$$

Therefore, we obtain the transformation rules as (for simplicity, we donot distinguish the symbols for matrix $\left[\kappa_f\right]$ and the tensor κ_f here as we start from and come back at Cartesian coordinate system when performing geometric transformations)

$$\begin{cases} \rho'_f = \rho_f / \det \mathbf{J}, \\ C'_f = C_f, \\ \kappa'_f = \mathbf{J}\kappa_f \mathbf{J}^T / \det \mathbf{J}, \\ v' = \mathbf{J}v, \\ \alpha'_f = \mathbf{J}\alpha_f \mathbf{J}^T / \det\mathbf{J}. \end{cases} \quad (5.5)$$

Now the question is how to control the velocity and flow density of fluids? An answer is to use porous media as we did in transforming thermal convection (Chap. 3). The flow velocity is determined by Darcy's law and the conservation law of fluid momentum. Here we write all the governing equations as

$$\begin{cases} \partial(\rho_m C_m T)/\partial t + \nabla \cdot \left(-\kappa_m \cdot \nabla T + \rho_f C_f vT - 16/3\beta_m^{-1} n_m^2 \sigma T^3 \cdot \nabla T\right) = 0, \\ \partial(\phi\rho_f)/\partial t + \nabla \cdot \left(\rho_f v\right) = 0, \\ v = -\eta/\mu_f \cdot \nabla P, \\ \rho_m C_m = (1-\phi)\rho_s C_s + \phi\rho_f C_f, \\ \kappa_m = (1-\phi)\kappa_s + \phi\kappa_f, \\ \beta_m^{-1} n_m^2 = (1-\phi)\beta_s^{-1} n_s^2 + \phi\beta_f^{-1} n_f^2. \end{cases} \quad (5.6)$$

where ρ_m, C_m, κ_m, β_m, and n_m are effective density, heat capacity, thermal conductivity, the Rosseland mean attenuation coefficient, and relative refraction index of the porous media, respectively. Similarly, ρ_s, C_s, κ_s, β_s, and n_s are the density, heat capacity, thermal conductivity, Rosseland mean attenuation coefficient, and relative refraction index of the solid material in the porous media. In addition ϕ is the porosity, η is the permeability, and μ_f is the dynamic viscosity coefficient. Here we use a simple volume-average method to calculate the effective parameters of porous media. The form-invariance of the equations above has been proven in the previous chapters and we can directly write the complete transformation rules

$$\begin{cases} \left(\rho_f C_f\right)' = \rho_f C_f, \\ \kappa'_f = \kappa_f, \\ \left(\beta_f^{-1} n_f^2\right)' = \beta_f^{-1} n_f^2, \\ \mu'_f = \mu_f \\ \phi' = \phi/\det\mathbf{J}, \\ \eta' = \mathbf{J}\eta\mathbf{J}^T/\det\mathbf{J}, \\ (\rho_s C_s)' = [(1-\phi)/(\det\mathbf{J} - \phi)]\,\rho_s C_s, \\ \kappa'_s = \left(\mathbf{J}\kappa_m\mathbf{J}^T - \phi\kappa_f\right)/(\det\mathbf{J} - \phi), \\ \left(\beta_s^{-1} n_s^2\right)' = \left(\mathbf{J}\left(\beta_m^{-1} n_m^2\right)\mathbf{J}^T - \phi\beta_f^{-1} n_f^2\right)/(\det\mathbf{J} - \phi). \end{cases} \tag{5.7}$$

Notice that here $\mathbf{v}' = \mathbf{J}\mathbf{v}/\det\mathbf{J}$ and can be obtained automatically after transforming porosity and permeability. Also, we don't transform the properties of pure fluids. Equation (5.7) is our comprehensive framework to handle heat transfer. There are a total of seven situations:

(1) There is only conduction. When the velocity $\mathbf{v}$ is zero and there is no radiation, the equation of heat transfer is

$$\partial(\rho_m C_m T)/\partial t + \nabla \cdot (-\kappa_m \cdot \nabla T) = 0. \tag{5.8}$$

We get the model of transforming thermal conduction and we only need to transform the thermal conductivity and density (or specific heat) of solid

$$\begin{cases} \left(\rho_f C_f\right)' = \rho_f C_f, \\ \kappa'_f = \kappa_f, \\ (\rho_s C_s)' = \dfrac{\rho_m C_m}{(1-\phi')\det\mathbf{J}} - \dfrac{\phi' C_f}{1-\phi'}, \\ \kappa'_s = \dfrac{\mathbf{J}\kappa_m\mathbf{J}^T}{(1-\phi')\det\mathbf{J}} - \dfrac{\phi'\kappa_f}{1-\phi'}. \end{cases} \tag{5.9}$$

Here ϕ' is not determined and can be chosen flexibly. For example, $\phi' = \phi$.
(2) There is only radiation. The equation of heat transfer for conduction and radiation is

$$\partial(\rho_m C_m T)/\partial t + \nabla \cdot \left(-16/3\beta_m^{-1} n_m^2 \sigma T^3 \cdot \nabla T\right) = 0. \tag{5.10}$$

The transformation rules are

$$\begin{cases} \left(\rho_f C_f\right)' = \rho_f C_f, \\ \left(\beta_f^{-1} \boldsymbol{n}_f^2\right)' = \beta_f^{-1} \boldsymbol{n}_f^2, \\ (\rho_s C_s)' = \dfrac{\rho_m C_m}{(1 - \phi') \det \mathbf{J}} - \dfrac{\phi' C_f}{1 - \phi'}, \\ \left(\beta_s^{-1} \boldsymbol{n}_s^2\right)' = \dfrac{\mathbf{J}\beta_m^{-1} \boldsymbol{n}_m^2 \mathbf{J}^T}{(1 - \phi') \det \mathbf{J}} - \dfrac{\phi' \beta_f^{-1} \boldsymbol{n}_f^2}{1 - \phi'}. \end{cases} \tag{5.11}$$

Again, ϕ' can be chosen flexibly.

(3) There is only convection. The governing equations are

$$\begin{cases} \partial(\rho_m C_m T)/\partial t + \boldsymbol{\nabla} \cdot \left(\rho_f C_f \boldsymbol{v} T\right) = 0, \\ \partial(\phi \rho_f)/\partial t + \boldsymbol{\nabla} \cdot \left(\rho_f \boldsymbol{v}\right) = 0, \\ \boldsymbol{v} = -\eta/\mu_f \cdot \boldsymbol{\nabla} P. \end{cases} \tag{5.12}$$

The transformation rules

$$\begin{cases} \left(\rho_f C_f\right)' = \rho_f C_f, \\ \mu_f' = \mu_f \\ \phi' = \phi/\det\mathbf{J}, \\ \eta' = \mathbf{J}\eta\mathbf{J}^T/\det\mathbf{J}, \\ (\rho_s C_s)' = [(1 - \phi) / (\det\mathbf{J} - \phi)] \rho_s C_s. \end{cases} \tag{5.13}$$

(4) There are only conduction and radiation. When the velocity $\boldsymbol{v}$ is zero, the equation of heat transfer for conduction and radiation is

$$\partial(\rho_m C_m T)/\partial t + \boldsymbol{\nabla} \cdot \left(-\kappa_m \cdot \boldsymbol{\nabla} T - 16/3\beta_m^{-1} \boldsymbol{n}_m^2 \sigma T^3 \cdot \boldsymbol{\nabla} T\right) = 0. \tag{5.14}$$

The transformation rules are

$$\begin{cases} \left(\rho_f C_f\right)' = \rho_f C_f, \\ \kappa_f' = \kappa_f, \\ \left(\beta_f^{-1} \boldsymbol{n}_f^2\right)' = \beta_f^{-1} \boldsymbol{n}_f^2, \\ (\rho_s C_s)' = \dfrac{\rho_m C_m}{(1 - \phi') \det \mathbf{J}} - \dfrac{\phi' C_f}{1 - \phi'}, \\ \kappa_s' = \dfrac{\mathbf{J}\kappa_m \mathbf{J}^T}{(1 - \phi') \det \mathbf{J}} - \dfrac{\phi' \kappa_f}{1 - \phi'}, \\ \left(\beta_s^{-1} \boldsymbol{n}_s^2\right)' = \dfrac{\mathbf{J}\beta_m^{-1} \boldsymbol{n}_m^2 \mathbf{J}^T}{(1 - \phi') \det \mathbf{J}} - \dfrac{\phi' \beta_f^{-1} \boldsymbol{n}_f^2}{1 - \phi'}. \end{cases} \tag{5.15}$$

Again, ϕ' can be chosen flexibly.

(5) There are only conduction and convection. The governing equations are

$$\begin{cases} \partial(\rho_m C_m T)/\partial t + \nabla \cdot \left(-\boldsymbol{\kappa}_m \cdot \nabla T + \rho_f C_f \boldsymbol{v} T\right) = 0, \\ \partial(\phi \rho_f)/\partial t + \nabla \cdot \left(\rho_f \boldsymbol{v}\right) = 0, \\ \boldsymbol{v} = -\boldsymbol{\eta}/\mu_f \cdot \nabla P. \end{cases} \quad (5.16)$$

The transformation rules

$$\begin{cases} \left(\rho_f C_f\right)' = \rho_f C_f, \\ \boldsymbol{\kappa}'_f = \boldsymbol{\kappa}_f, \\ \mu'_f = \mu_f \\ \phi' = \phi/\det \mathbf{J}, \\ \boldsymbol{\eta}' = \mathbf{J}\boldsymbol{\eta}\mathbf{J}^\mathsf{T}/\det \mathbf{J}, \\ \left(\rho_s C_s\right)' = [(1-\phi) / (\det \mathbf{J} - \phi)]\rho_s C_s, \\ \boldsymbol{\kappa}'_s = \left(\mathbf{J}\boldsymbol{\kappa}_m \mathbf{J}^\mathsf{T} - \phi \boldsymbol{\kappa}_f\right) / (\det \mathbf{J} - \phi). \end{cases} \quad (5.17)$$

When convection exists, ϕ' is determined by transformation theory.

(6) There are only convection and radiation. The governing equations are

$$\begin{cases} \partial(\rho_m C_m T)/\partial t + \nabla \cdot \left(\rho_f C_f \boldsymbol{v} T - 16/3\beta_m^{-1}\boldsymbol{n}_m^2 \sigma T^3 \cdot \nabla T\right) = 0, \\ \partial(\phi \rho_f)/\partial t + \nabla \cdot \left(\rho_f \boldsymbol{v}\right) = 0, \\ \boldsymbol{v} = -\boldsymbol{\eta}/\mu_f \cdot \nabla P. \end{cases} \quad (5.18)$$

The transformation rules

$$\begin{cases} \left(\rho_f C_f\right)' = \rho_f C_f, \\ \left(\beta_f^{-1}\boldsymbol{n}_f^2\right)' = \beta_f^{-1}\boldsymbol{n}_f^2, \\ \mu'_f = \mu_f \\ \phi' = \phi/\det \mathbf{J}, \\ \boldsymbol{\eta}' = \mathbf{J}\boldsymbol{\eta}\mathbf{J}^\mathsf{T}/\det \mathbf{J}, \\ \left(\rho_s C_s\right)' = [(1-\phi) / (\det \mathbf{J} - \phi)]\rho_s C_s, \\ \left(\beta_s^{-1}\boldsymbol{n}_s^2\right)' = \left(\mathbf{J}\left(\beta_m^{-1}\boldsymbol{n}_m^2\right)\mathbf{J}^\mathsf{T} - \phi\beta_f^{-1}\boldsymbol{n}_f^2\right) / (\det \mathbf{J} - \phi). \end{cases} \quad (5.19)$$

(7) The most general case where conduction, convection and radiation all exist. This is just the story Eqs. (5.6) and (5.7) tell.

5.2 Applications

Again, we can check the transformation theory by considering some typical functions. The space transformations of cloak, concentrator, and rotator can be, respectively, expressed as

$$\begin{cases} r' = (R_2 - R_1)\, r/R_2 + R_1, \\ \theta' = \theta, \end{cases} \quad (5.20)$$

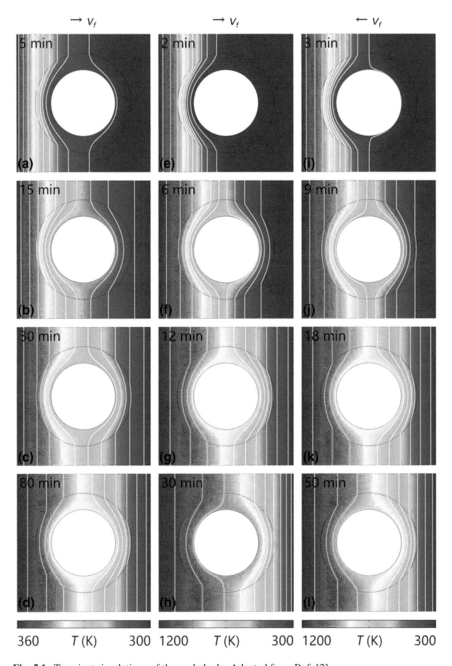

Fig. 5.1 Transient simulations of thermal cloak. Adapted from Ref. [2]

$$\begin{cases} r' = R_1 r/R_m \text{ with } r < R_m, \\ r' = ((R_1 - R_m) R_2 + (R_2 - R_1) r) / (R_2 - R_m) \text{ with } R_m < r < R_2, \\ \theta' = \theta, \end{cases} \quad (5.21)$$

$$\begin{cases} r' = r \\ \theta' = \theta + \theta_0 \text{ with } r < R_1, \\ \theta' = \theta + \theta_0 (r - R_2) / (R_1 - R_2) \text{ with } R_1 < r < R_2, \end{cases} \quad (5.22)$$

where R_1 and R_2 are the inner and outer radiuses, R_m is a medium value determining the concentrating ratio, and θ_0 is rotation degree. Then, the Jacobian transformation matrix of Eqs. (5.20)–(5.22) can be derived as

$$\mathbf{J} = \begin{pmatrix} \partial r'/\partial r & \partial r'/(r\partial\theta) \\ r'\partial\theta'/\partial r & r'\partial\theta'/(r\partial\theta) \end{pmatrix}. \quad (5.23)$$

We show the numerical results of cloaks in Fig. 5.1, where white lines represent isotherms. The size of each simulation is 0.1×0.1 m², $R_1 = 0.024$ m, and $R_2 = 0.036$ m; and the background fluid has parameters $\rho_f = 10^3$ kg m^{-3}, $C_f = 10^3$ J kg^{-1}K^{-1}, $\kappa_f = 1$ Wm^{-1}K^{-1}, $v_f = 10^{-5}$ m s^{-1}, $\beta_f = 100$ m^{-1}, and $n_f = 1$ throughout this chapter. The transformation media and velocity are designed according to Eq. (5.20). (a)–(d) show the temperature evolutions over time with a temperature interval 300–360 K and a background fluid velocity along the $+x$ axis. (e)–(h) show the temperature evolutions over time with a temperature interval 300–1200 K and a background fluid velocity along the $+x$ axis. (i)–(l) show the temperature evolutions over time with a temperature interval 300–1200 K and a background fluid velocity along the $-x$ axis. Following Eq. (5.20) (which yields Fig. 5.1), we can easily give the transient simulations of thermal concentrator and thermal rotator according to Eq. (5.21) and Eq. (5.22), respectively, which are not shown in this chapter.

5.3   Exercises and Solutions

Exercises

1. Prove that the choice of reference temperature shall not change the validity and form-invariance of Eq. (5.1).

Solutions

1. **Solution**: Rewrite the left-hand side of Eq. (5.1) as $\partial(\rho_f C_f (T - T_{Ref}))/\partial t + \nabla \cdot \left(-\kappa_f \cdot \nabla T + \rho_f C_f v(T - T_{Ref}) - 16/3\beta_f^{-1} n_f^2 \sigma T^3 \cdot \nabla T\right)$. We only need to consider the term

$$\partial(\rho_f C_f T_{Ref})/\partial t + \nabla \cdot (\rho_f C_f v T_{Ref}) = T_{Ref}(\partial(\rho_f C_f)/\partial t + \nabla \cdot (\rho_f C_f v)). \quad (5.24)$$

Notice the conservation law of momentum in pure fluids

$$\partial \rho_f / \partial t + \nabla \cdot \left(\rho_f \boldsymbol{v} \right) = 0. \tag{5.25}$$

Since C_f is a constant, we can see $\partial(\rho_f C_f T_{Ref})/\partial t + \nabla \cdot \left(\rho_f C_f \boldsymbol{v} T_{Ref} \right) = 0$ and Eq. (5.1) is still valid and form-invariant. Then we obtain the same conclusion for Eq. (5.6).

References

1. Rosseland, S.: Theoretical Astrophysics. Oxford University Press, Clarendon, London and New York (1936)
2. Xu, L.J., Yang, S., Dai, G.L., Huang, J.P.: Transformation omnithermotics: simultaneous manipulation of three basic modes of heat transfer (Submitted) (2019)

Chapter 6
Macroscopic Theory for Thermal Composites: Effective Medium Theory, Rayleigh Method and Perturbation Method

Abstract Thermal conductivity can depend on temperature (namely, nonlinear), which is common in nature. Since composites are widely used in thermal metamaterials to tailor thermal conductivities and other macroscopic properties, a fundamental problem is how to predict the effective thermal conductivity (κ_e) of composites. In this chapter, we present various kinds of theories or methods to calculate both linear and nonlinear part of κ_e, which include the effective medium theory, the Rayleigh method and the perturbation method. We show their validity by comparing with the numerical results from finite-element simulations for periodic composites with linear or nonlinear thermal conductivities. Also, we investigate the condition for generating nonlinearity enhancement.

Keywords Composites · Nonlinear conduction · Effective medium approximation · The Rayleigh method · Perturbation method

6.1 Linear Part of Effective Thermal Conductivity

In this section we will give two methods, effective medium theory and the Rayleigh method, to calculate κ_e for linear media, which means the thermal conductivities are constant under different temperatures. First, we define the effective thermal conductivity κ_e of a thermal-diffusion system based on

$$\langle \boldsymbol{J} \rangle = -\kappa_e \langle \nabla T \rangle \tag{6.1}$$

In this equation, κ_e is generally a tensor since the volume-average heat flux $\langle \boldsymbol{J} \rangle$ and volume-average temperature gradient $\langle \nabla T \rangle$ are vectors. Usually we focus on the conductivity on a given direction such as the x direction, we can write

$$\kappa_e = -\langle \boldsymbol{J}_x \rangle / \langle \nabla T_x \rangle. \tag{6.2}$$

For simplicity, we shall mainly introduce the cases of binary composites in two dimensions, and denote the thermal conductivity as κ_i and κ_h for inclusions and the host, respectively.

© Springer Nature Singapore Pte Ltd. 2020
J.-P. Huang, *Theoretical Thermotics*,
https://doi.org/10.1007/978-981-15-2301-4_6

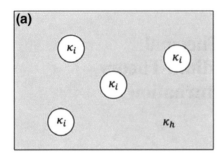
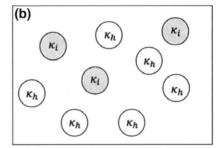

Fig. 6.1 Schematic diagram of typical composites for **a** Maxwell-Garnett formula and **b** Bruggeman formula. Adapted from Ref. [3]

6.1.1 Effective Medium Theory

The effective medium theory gives an analytical model to predict the macroscopic properties of composites approximately, in which the volume or area fraction of the inclusions f_i (thus the area fraction of the host satisfies $f_h = 1 - f_i$) plays an important role. There are two basic models of effective medium theory, the Maxwell-Garnett [1] formula and the Bruggeman formula [2]. We give the results for two-component composite and extensions to multi-component composite can be straightforward.

Maxwell-Garnett Formula

Figure 6.1a illustrates a typical composite which can be described by Maxwell-Garnett formula, where circular inclusions are embedded randomly in the host. The Maxwell-Garnett formula for two-dimensional two-component composites is solved from

$$\frac{\kappa_e - \kappa_h}{\kappa_e + \kappa_h} = f_i \frac{\kappa_i - \kappa_h}{\kappa_i + \kappa_h}. \tag{6.3}$$

For three-dimensional cases, the Maxwell-Garnett formula writes

$$\frac{\kappa_e - \kappa_h}{\kappa_e + 2\kappa_h} = f_i \frac{\kappa_i - \kappa_h}{\kappa_i + 2\kappa_h}. \tag{6.4}$$

Now we will give a brief proof of Maxwell-Garnett formula for two-dimensional case. According to the definition of κ_e, we have

$$\begin{aligned}
\kappa_e &= \frac{f_i \langle \boldsymbol{J}_i \rangle + (1 - f_i) \langle \boldsymbol{J}_h \rangle}{f_i \langle \nabla T_i \rangle + (1 - f_i) \langle \nabla T_h \rangle} \\
&= \frac{f_i \kappa_i \langle \nabla T_i \rangle + (1 - f_i) \kappa_h \langle \nabla T_h \rangle}{f_i \langle \nabla T_i \rangle + (1 - f_i) \langle \nabla T_h \rangle},
\end{aligned} \tag{6.5}$$

where $\langle J_j \rangle, \langle \nabla T_j \rangle$ are the volume/area-average heat flux and temperature and the notes $j = i, h$ represent the inclusions and host respectively. The Fourier law for heat conduction for steady states writes

$$\nabla \cdot \left(\kappa_j \nabla T_j \right) = 0, \quad j = i, h. \tag{6.6}$$

Now we assume that the inclusions are so dilute that we can only find one inclusion or particle in an infinite (two-dimensional) plane. In addition, the inclusions are circular. With the boundary conditions (ξ denotes the boundary between the inclusions and the host) in polar coordinates (r, θ)

$$\begin{cases} T_i(r = 0) \text{ is finite,} \\ T_h(r \to \infty) = \nabla T_h((r \to \infty)) \cdot e_x \\ T_i(\xi) = T_h(\xi), \\ -\kappa_i \left. \frac{\partial T}{\partial \xi} \right|_\xi = -\kappa_h \left. \frac{\partial T_2}{\partial \xi} \right|_\xi \end{cases} \tag{6.7}$$

we can find [4]

$$\langle \nabla T_i \rangle = \frac{2\kappa_h}{\kappa_i + \kappa_h} \langle \nabla T_h \rangle . \tag{6.8}$$

Finally we obtain the expression for κ_e

$$\kappa_e = \kappa_h \frac{\kappa_h(1 - f_i) + \kappa_i(1 + f_i)}{\kappa_h(1 + f_i) + \kappa_i(1 - f_i)}, \tag{6.9}$$

which is consistent with the Maxwell-Garnett formula.

Bruggeman Formula

The Bruggeman formula for two-dimensional binary composites is given by

$$f_i \frac{\kappa_e - \kappa_i}{\kappa_e + \kappa_i} + f_h \frac{\kappa_e - \kappa_h}{\kappa_e + \kappa_h} = 0. \tag{6.10}$$

Also, for the three-dimensional case, the Bruggeman formula writes

$$f_i \frac{\kappa_e - \kappa_i}{2\kappa_e + \kappa_i} + f_h \frac{\kappa_e - \kappa_h}{2\kappa_e + \kappa_h} = 0. \tag{6.11}$$

Fig. 6.1b illustrates a typical composite that can be described by the Bruggeman formula. One difference between the Bruggeman and Maxwell-Garnett formula is that the two components for the Bruggeman formula are symmetrical. In fact, the indexes "*i*" and "*h*" can be exchanged and κ_e will not be influenced.

6.1.2 The Rayleigh Method

The Rayleigh method [5] is a first-principle approach to predict effective conductivity especially for periodic composites, and it directly solves the heat conduction equation with certain boundary conditions. To distinguish from the following nonlinear cases, we denote the temperature distribution as $T^{(0)}$ in the linear case. The general solution of Eq. (6.6) for the area occupied by the inclusions can be written in polar coordinates (r, θ) as

$$T_i^{(0)}(\rho, \theta) = C_{00} + \sum_{m=1}^{\infty} C_{0m}^2 r^m \cos(m\theta) + C_{0m}^1 r^m \sin(m\theta), \tag{6.12}$$

where the corresponding polar coordinate system has a pole at the core of a selected cell in the composite. For the host, we can also write

$$T_h^{(0)}(\rho, \theta) = A_{00} + \sum_{m=1}^{\infty} (A_{0m}^2 r^m + B_{0m}^2 r^{-m}) \cos(m\theta) + (A_{0m}^1 r^m + B_{0m}^1 r^{-m}) \sin(m\theta). \tag{6.13}$$

To find the coefficients (A_{00}, A_{0m}^2, A_{0m}^1, B_{0m}^1, B_{0m}^2, C_{00}, C_{0m}^1 and C_{0m}^2), we should apply both the boundary conditions between the inclusions and host (boundaries denoted as ξ)

$$T_i^{(0)} = T_h^{(0)}|_\xi, \tag{6.14a}$$

$$\kappa_i \frac{\partial T_i^{(0)}}{\partial r} = \kappa_h \frac{\partial T_h^{(0)}}{\partial r}\bigg|_\xi, \tag{6.14b}$$

and the Rayleigh identity for periodic composites [6]

$$A_{00} + \sum_{m=1}^{\infty} \rho^m [A_{0m}^2 \cos(m\theta) + A_{0m}^1 \sin(m\theta)]$$

$$= \sum_{k=1}^{\infty} \sum_{m=1}^{\infty} \rho_k^{-m} [B_{0m}^2 \cos(m\theta_k) + B_{0m}^1 \sin(m\theta_k)] - \frac{T_L - T_R}{L}, \tag{6.15}$$

where

$$\rho_k = \sqrt{(x - u_k)^2 + (y - v_k)^2}, \tag{6.16a}$$

$$\cos \theta_k = (x - u_k)/r_k. \tag{6.16b}$$

Here u_k and v_k are the Cartesian coordinates of the central point in the k-th cell. k is positive and we take the cell (where the pole lies in) as the 0-th one. Again we set the external temperature gradient to be along the x direction. We can notice that, when only one particle is embedded, the boundary condition at infinity (see Eq. (6.7)) is often used instead of the Rayleigh identity. By partially differentiating with respect to x in both sides of point Q and taking the value at Q (which can be arbitrary with Cartesian or polar coordinates being (x_0, y_0) or (r_0, θ_0)), the Rayleigh identity tells [6]

$$\sum_{m=1}^{\infty} \frac{m! r_0^{m-n}(\kappa_h + \kappa_i)}{(m-n)!(\kappa_h - \kappa_i)} [A_{0m}^2 \cos((m-n)\theta_0) + A_{0m}^1 \sin((m-n)\theta_0)] - \sum_{m=1}^{\infty}(-1)^n \frac{(m+n-1)!}{(m-1)!}$$

$$\times [B_{0m}^2 W_{m+n}^2(Q) + B_{0m}^1 W_{m+n}^1(Q)] = -\frac{T_L - T_R}{L}\delta_{1,n}$$

$$(6.17)$$

where

$$W_l^1(Q) = \sum_{k=1}^{\infty} \rho_k^{-l} \sin(l\theta_k), \tag{6.18a}$$

$$W_l^2(Q) = \sum_{k=1}^{\infty} \rho_k^{-l} \cos(l\theta_k). \tag{6.18b}$$

By truncating the series expansions [since we assume that the temperature gradient is applied along the x direction, we only remain A_{01}^2, A_{03}^2, B_{01}^2 and B_{03}^2 in Eq. (6.17)], we can write the approximate solutions in the selected cell as

$$T_i^{(0)}(\rho, \theta) = C_{00} + C_{01}^2 r \cos\theta + C_{03}^2 r^3 \cos(3\theta) \tag{6.19}$$

and

$$T_h^{(0)}(\rho, \theta) = C_{00} + A_{01}^2 r \cos\theta + A_{03}^2 r^3 \cos(3\theta) + B_{01}^2 r^{-1} \cos\theta + B_{03}^2 r^{-3} \cos(3\theta), \tag{6.20}$$

where [6]

$$B_{01}^2 = -\frac{T_L - T_R}{L} \Big/ \Big(\frac{\kappa_h + \kappa_i}{a^2(\kappa_h - \kappa_i)} - \frac{3a^6(W_4^2)^2(\kappa_h - \kappa_i)}{\kappa_h + \kappa_i}\Big), \tag{6.21a}$$

$$B_{03}^2 = -\frac{a^6 W_4^2(\kappa_h - \kappa_i)}{\kappa_h + \kappa_i}) B_{01}^2, \tag{6.21b}$$

$$A_{0m}^2 = \frac{\kappa_h + \kappa_i}{a^{2m}(\kappa_h - \kappa_i)} B_{0m}^2, \tag{6.21c}$$

$$C_{0m}^2 = \frac{2\kappa_h}{a^{2m}(\kappa_h - \kappa_i)} B_{0m}^2, \tag{6.21d}$$

$$W_4^2 = 3.13085/S^2. \tag{6.21e}$$

Here a is the radius of a single particle in the unit cell with area S. The only difference from its electrical counterpart is that C_{00} is usually neglected in electricity because a constant term of electrical potentials should not change any corresponding physical properties. Fortunately, in the linear case, C_{00} (though still to be determined by more conditions) should not influence the result of effective conduction either. Also, $W_l^2(Q)$ is position-independent.

The effective linear thermal conductivity κ_e can be calculated by [7, 8]

$$\kappa_e = \kappa_h + (\kappa_i - \kappa_h)\langle \nabla_x T^{(0)} \rangle_i / \langle \nabla_x T^{(0)} \rangle. \tag{6.22}$$

In fact Eq. (6.22) is another form of Eq. (6.2). Also it is easy to see

$$\langle \nabla_x T^{(0)} \rangle_i = C_{01}^2 f_i, \tag{6.23}$$

$$f_i = \pi a^2 / S. \tag{6.24}$$

Note there exists [6]

$$\langle \nabla_x T^{(0)} \rangle = \frac{2\kappa_h}{\kappa_h + \kappa_e} \frac{T_R - T_L}{L}, \tag{6.25}$$

which is consistent with the Bruggeman theory. Then we can find that κ_e has the same form as its electrical counterpart in the case of linear conduction, namely,

$$\kappa_e = \kappa_h \frac{(-\beta_1 + \beta_1 f_i + f_i^4)\kappa_h^2 - 2(\beta_1 + f_i^4)\kappa_h\kappa_i + (-\beta_1 - \beta_1 f_i + f_i^4)\kappa_i^2}{(-\beta_1 - \beta_1 f_i + f_i^4)\kappa_h^2 - 2(\beta_1 + f_i^4)\kappa_h\kappa_i + (-\beta_1 + \beta_1 f_i + f_i^4)\kappa_i^2}, \tag{6.26}$$

where $\beta_1 = \frac{\pi^4}{3(W_4^2 S^2)^2} = 3.31248$.

Now we turn back to the Rayleigh identity Eq. (6.15). What's physical figure? The temperature distribution in the host originates from spatial infinity (the heat sources on the boundary), host and inclusions, which is another expression different from the general solution given by Eq. (6.13). If there are no inclusions, the infinity and host correspond to a temperature distribution proportional to $-x$ if the temperature gradient is set along the x direction. Since the host and inclusions have different conductivities, the inclusions correspond to terms decreasing when the distance of host and an inclusion (denoted as ρ_k) increases. So we can write

$$T_0^h = \sum_{m=1}^{\infty} \rho^{-m} [B_{0m}^2 \cos(m\theta) + B_{0m}^1 \sin(m\theta)]$$

$$+ \sum_{k=1}^{\infty} \sum_{m=1}^{\infty} \rho_k^{-m} [B_{0m}^2 \cos(m\theta_k) + B_{0m}^1 \sin(m\theta_k)] - \frac{T_L - T_R}{L} x. \tag{6.27}$$

The term $\frac{T_R - T_L}{L} x$ corresponds to infinity while the term $\sum_{m=1}^{\infty} \rho^{-m} [B_{0m}^2 \cos(m\theta) + B_{0m}^1 \sin(m\theta)]$ corresponds to the inclusion in the 0-th cell. The terms $\sum_{m=1}^{\infty} \rho_k^{-m} [B_{0m}^2 \cos(m\theta_k) + B_{0m}^1 \sin(m\theta_k)]$ come from the periodicity and represent the influence from the k-th cell. Because the two expressions Eqs. (6.13) and (6.27) must give the same result, the Rayleigh identity can be obtained.

6.2   Nonlinear Part of Effective Thermal Conductivity

In this section, we further take temperature-dependent thermal conductivity $\kappa(T)$ into consideration. Again, for a two-component composite, either the inclusion or the host can have a nonlinear thermal conductivity now. In the following discussion, for simplicity, we assume that either the inclusion or host can be nonlinear with a thermal conductivity dependent on temperature (T) whose κ_j can be written as

$$\kappa_j = \kappa_{j0} + \chi_j (T + T_{rt})^\alpha \quad (j = i, h). \tag{6.28}$$

Here κ_{j0} is the linear (namely, temperature-independent) part of κ_j, χ_j is the non-linearity (temperature-dependence) coefficient, T_{rt} is the reference temperature for measuring nonlinearity and α can be any real number. For weak nonlinearity meaning $\chi_j (T + T_{rt})^\alpha \ll \kappa_{j0}$, the effective conductivity can be written as

$$\kappa_e = \kappa_{e0} + \chi_e (T + T_{rt})^\alpha + o[(T + T_{rt})^{2\alpha}], \tag{6.29}$$

where κ_{e0} is the linear solution which has been discussed in the previous section, χ_e is the effective nonlinearity coefficient, and $o[(T + T_{rt})^{2\alpha}]$ denotes higher-order terms [which are much smaller than $\chi_e (T + T_{rt})^\alpha$]. To see the enhancement or reduction of nonlinearity, we consider two cases where only the inclusions or the host is nonlinear. We define the nonlinearity enhancement ratio as

$$c = \frac{\chi_e}{\chi_j} \tag{6.30}$$

where $j = i$ for the first case (i.e., only the inclusions are nonlinear: $\chi_i \neq 0$ and $\chi_h = 0$) while $j = h$ for the second (namely, only the host is nonlinear: $\chi_h \neq 0$ and $\chi_i = 0$). Also, Fig. 6.2 illustrates the schematic diagram for the two cases, where (a) denotes the first and (b) denotes the second one.

6.2.1   Effective Medium Theory

Effective medium theory can be applied to nonlinear cases if we combine series expansion with it. Cutting off the Taylor's series expansions of Eq. (6.29), we can find that

$$\chi_e (T + T_{rt})^\alpha = \frac{\partial \kappa_e}{\partial \chi_i} \tag{6.31}$$

where κ_e is a function of κ_{j0}, f_j and χ_j.

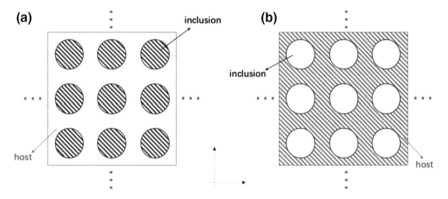

Fig. 6.2 Schematic diagrams illustrating two classes of periodic composites: the circles represent the inclusions and the shadow lines represent the nonlinear material. **a** nonlinear inclusions are periodically embedded in a linear host; **b** linear inclusions are periodically embedded in a nonlinear host. Adapted from Ref. [3]

Maxwell-Garnett Formula

For the first case, only the inclusion is nonlinear and then the nonlinearity coefficient ratio $c\ (= \chi_e/\chi_i)$ is given by

$$c = \frac{4 f_i}{\left(1 + \dfrac{\kappa_{i0}}{\kappa_{h0}} + f_i - f_i \dfrac{\kappa_{i0}}{\kappa_{h0}}\right)^2}. \tag{6.32}$$

For the second case where only the host is nonlinear, we can also obtain the nonlinearity coefficient ratio $c\ (= \chi_e/\chi_h)$ as

$$c = \frac{(1 - f_i^2)\left[1 + \left(\dfrac{\kappa_{i0}}{\kappa_{h0}}\right)^2\right] + 2(1 - f_i)^2 \dfrac{\kappa_{i0}}{\kappa_{h0}}}{\left(1 + \dfrac{\kappa_{i0}}{\kappa_{h0}} + f_i - f_i \dfrac{\kappa_{i0}}{\kappa_{h0}}\right)^2}. \tag{6.33}$$

Bruggeman Formula

Similarly, based on the Bruggeman formula, the nonlinearity coefficient ratio $c\ (= \chi_e/\chi_i)$ for the first case is

$$c = \frac{1}{2}\left[\frac{(2 f_i - 1)\left(2 f_i - 2 f_i \dfrac{\kappa_{h0}}{\kappa_{i0}} - 1 + \dfrac{\kappa_{h0}}{\kappa_{i0}}\right) + 2 \dfrac{\kappa_{h0}}{\kappa_{i0}}}{\sqrt{\left(2 f_i - 2 f_i \dfrac{\kappa_{h0}}{\kappa_{i0}} - 1 + \dfrac{\kappa_{h0}}{\kappa_{i0}}\right)^2 + 4 \dfrac{\kappa_{h0}}{\kappa_{i0}}}} + 2 f_i - 1\right] \tag{6.34}$$

and the nonlinearity coefficient ratio $c \; (= \chi_e/\chi_h)$ for the second case is

$$
c = \frac{1}{2} \left[\frac{(2f_i - 1)\left(2f_i - 2f_i\dfrac{\kappa_{i0}}{\kappa_{h0}} - 1 + \dfrac{\kappa_{i0}}{\kappa_{h0}}\right) + 2\dfrac{\kappa_{i0}}{\kappa_{h0}}}{\sqrt{\left(2f_i - 2f_i\dfrac{\kappa_{i0}}{\kappa_{h0}} - 1 + \dfrac{\kappa_{i0}}{\kappa_{h0}}\right)^2 + 4\dfrac{\kappa_{i0}}{\kappa_{h0}}}} - 2f_i + 1 \right]. \qquad (6.35)
$$

6.2.2 The Rayleigh Method

For the nonlinear conduction, the thermal conductivity depends on the temperature (i.e., potential) while the electrical conductivity relies on the field (namely, the gradient of potential). This difference makes it much more complicated if we still use Rayleigh identity (for nonlinear conduction) combined with perturbation theory to solve a nonhomogeneous equation (we will give a simple example of perturbation theory later) following the framework in electrical conduction. Nevertheless, we can take Eq. (6.28) into Eq. (6.26) and still use Taylor series to calculate c through

$$
c = \frac{\partial \kappa_e}{\chi_j (T + T_{\mathrm{rt}})^\alpha \partial \chi_j}. \qquad (6.36)
$$

Finally, the expressions of c for the two cases we study are

$$
c = \frac{4\beta_1 f_i \kappa_{h0}^2 \left[\beta_1 (\kappa_{h0} + \kappa_{i0})^2 + f_i^4 (\kappa_{h0} - \kappa_{i0})^2 \right]}{\left[\beta_1 (\kappa_{h0} + \kappa_{i0})(f_i \kappa_{h0} - f_i \kappa_{i0} + \kappa_{h0} + \kappa_{i0}) - f_i^4 (\kappa_{h0} - \kappa_{i0})^2 \right]^2} \qquad (6.37)
$$

for the first case and

$$
\begin{aligned}
c = {} & \frac{-\beta_1^2 (f_i - 1)(\kappa_{h0} + \kappa_{i0})^2 \left[(f_i + 1)\kappa_{h0}^2 - 2(f_i - 1)\kappa_{h0}\kappa_{i0} + (f_i + 1)\kappa_{i0}^2 \right]}{\left[\beta_1 (\kappa_{h0} + \kappa_{i0})(f_i \kappa_{h0} - f_i \kappa_{i0} + \kappa_{h0} + \kappa_{i0}) - f_i^4 (\kappa_{h0} - \kappa_{i0})^2 \right]^2} \\
& + \frac{-2\beta_1 f_i^4 (\kappa_{h0} - \kappa_{i0})^2 \left[2(f_i + 1)\kappa_{h0}\kappa_{i0} + \kappa_{h0}^2 + \kappa_{i0}^2 \right] + f_i^8 (\kappa_{h0} - \kappa_{i0})^4}{\left[\beta_1 (\kappa_{h0} + \kappa_{i0})(f_i \kappa_{h0} - f_i \kappa_{i0} + \kappa_{h0} + \kappa_{i0}) - f_i^4 (\kappa_{h0} - \kappa_{i0})^2 \right]^2}
\end{aligned}
$$
$$(6.38)$$

for the second case. It can be easily confirmed that c only depends on area fraction f_i and ratio κ_{i0}/κ_{h0}. In addition, the expressions of c keep the same for different α and T_{rt}, two additional parameters as adopted in Eq. (6.28).

6.2.3 The Perturbation Method

The perturbation theory is another effective method to deal with nonlinear systems such as those in nonlinear optics [7, 8]. Reference [9] gives an example about how to deal with nonlinear heat conduction by using the perturbation theory. Again we consider a weak nonlinear relationship which can be generally expressed as

$$\kappa = \kappa_0 + \kappa_1 T + \kappa_2 T^2 + \kappa_3 T^3 + \cdots \tag{6.39}$$

where κ_n is the expansion parameters. To apply the perturbation theory, we should always have the following condition,

$$|\kappa_0| \gg |\kappa_1 T| \gg |\kappa_2 T^2| \gg |\kappa_3 T^3| \gg \cdots \tag{6.40}$$

The temperature itself can also be expanded as

$$T = T^{(0)} + T^{(1)} + T^{(2)} + T^{(3)} + \cdots \tag{6.41}$$

Similarly, the expansion of heat flux $\boldsymbol{J}$ is

$$\boldsymbol{J} = -(\kappa_0 + \kappa_1 T + \kappa_2 T^2 + \cdots)\nabla(T^{(0)} + T^{(1)} + T^{(2)} + \cdots). \tag{6.42}$$

Now we write the first two terms of heat flux as

$$\boldsymbol{J}^{(0)} = -\kappa_0 \nabla T^{(0)}, \tag{6.43}$$

and

$$\boldsymbol{J}^{(1)} = -\kappa_0 \nabla T^{(1)} - \kappa_1 T^{(0)} \nabla T^{(0)}. \tag{6.44}$$

Using the above expansions, we can also obtain conduction equations for $T^{(n)}$:

$$\rho c \frac{\partial T^{(0)}}{\partial t} = \nabla \cdot (\kappa_0 \nabla T^{(0)}), \tag{6.45}$$

$$\rho c \frac{\partial T^{(1)}}{\partial t} = \nabla \cdot (\kappa_0 \nabla T^{(1)} + \kappa_1 T^{(0)} \nabla T^{(0)}), \tag{6.46}$$

The simplest case is embedding a nonlinear circular inclusion (with its radius denoted as r_i) into a linear host medium and we still write $\kappa_i = \kappa_{i0} + \chi_i T$. The boundary condition at $r = r_i$ is

$$\begin{cases} T_h^{(k)} = T_i^{(k)} \\ \boldsymbol{J}_h^{(k)} \cdot \hat{\boldsymbol{n}} = \boldsymbol{J}_i^{(k)} \cdot \hat{\boldsymbol{n}} \end{cases}, \tag{6.47}$$

where $\hat{n}$ is the unit normal vector on the surface of the inclusion. In addition, the boundary condition at $r = \infty$ and $r = 0$ is

$$\begin{cases} \nabla T_h(r = \infty) = H e_x \\ |T_i^{(k)}(r = 0)| < \infty \end{cases}. \tag{6.48}$$

Here T_i and J_i denote temperature and heat flux density of inclusion respectively and T_h and J_h represent those of host medium. In addition, we take H as a given number which can be seen as the outer temperature gradient applied on the composite.

It is easy to solve out the 0-order solution,

$$\begin{cases} T_h^{(0)} = \dfrac{G}{r} \cos\theta + Hr \cos\theta \\ T_i^{(0)} = Cr \cos\theta \end{cases}, \tag{6.49}$$

where the coefficients are

$$\begin{cases} G = Hr_i^2 \dfrac{\kappa_{h0} - \kappa_{i0}}{\kappa_{h0} + \kappa_{i0}} \\ C = 2H \dfrac{\kappa_{h0}}{\kappa_{h0} + \kappa_{i0}} \end{cases}. \tag{6.50}$$

Here a constant term of temperature can be neglected because this will not change the results below (for zero-order and 1st-order). With 0-order solution, the 1st-order equation is solved out as

$$\nabla^2 T_i^{(1)} = -\frac{\chi_i}{\kappa_{i0}} C^2. \tag{6.51}$$

To solve such a non-homogeneous equation, we first find a special solution

$$-\frac{\chi_i}{4\kappa_{i0}} C^2 r^2 \tag{6.52}$$

and then set the general solution as

$$\begin{cases} T_h^{(1)} = \dfrac{D}{r^2} \cos 2\theta \\ T_i^{(1)} = Er^2 \cos 2\theta - \dfrac{\chi_i}{4\kappa_{i0}} C^2 r^2 + F \end{cases}. \tag{6.53}$$

Finally we can write all the coefficients as

$$\begin{cases} F = \dfrac{\chi_i}{4\kappa_{i0}} C^2 r_i^2 \\ D = -\dfrac{C^2}{4} \dfrac{\chi_i}{\kappa_{i0} + \kappa_{h0}} r_i^4 \\ E = -\dfrac{C^2}{4} \dfrac{\chi_i}{\kappa_{i0} + \kappa_{h0}} \end{cases}. \tag{6.54}$$

To calculate the effective nonlinear conductivity, we define a constitutive relationship between average heat flux and average thermal field as

$$\langle J \rangle = -\kappa_{e0} \langle \nabla T \rangle - \chi_e \langle T \rangle \langle \nabla T \rangle. \tag{6.55}$$

From Refs. [7, 8], we have (S denotes the area occupied by the composite)

$$\frac{1}{S} \int_S J - (-\kappa_{h0} \nabla T) \mathrm{d}\Omega = \langle J \rangle + \kappa_{h0} \langle \nabla T \rangle. \tag{6.56}$$

Using Eq. (6.55) to replace $\langle J \rangle$ in the right-hand side of Eq. (6.56), we get

$$\frac{1}{S} \int_{S_i} ((\kappa_{h0} - \kappa_{i0}) \nabla T - \chi_i T \nabla T) \mathrm{d}\Omega \tag{6.57}$$

$$= (\kappa_{h0} - \kappa_{e0}) \langle \nabla T \rangle - \chi_e \langle T \rangle \langle \nabla T \rangle.$$

Substituting Eq. (6.49) into Eq. (6.57), we have

$$\kappa_{e0} = \kappa_{h0} - 2 f_i \kappa_{h0} \frac{\kappa_{h0} - \kappa_{i0}}{\kappa_{h0} + \kappa_{i0}}, \tag{6.58}$$

noticing $f_i = \frac{\pi r_i^2}{S}$ is the area fraction of inclusion. Similarly, we can see

$$\chi_e = 2 f_i \chi_i \frac{\kappa_{h0}}{\kappa_{h0} + \kappa_{i0}}. \tag{6.59}$$

6.3 Examples

For example, we compare predictions of theories above with the simulation results from finite-element analysis. We set up a model where circular inclusions periodically embedded into the host. Again, as shown in Fig. 6.2, we consider two cases. In Fig. 6.2a only the inclusions are nonlinear while in Fig. 6.2b only the host is nonlinear. The size of the composite material is 20 cm × 20 cm and each unit cell is 1 cm × 1 cm. Also, two heat sources put on boundaries are set at $T_L = 313$ K and $T_R = 273$ K. If no overlapping exists, there is an upper limit of area fraction $f_i < \pi/4$ for circular inclusions. Without loss of generality, we take $\alpha = 1$ and $T_{\mathrm{rt}} = 0$ K in all simulations.

For weak nonlinearity, Figs. 6.3 and 6.4 respectively illustrate the effective linear thermal conductivity κ_{e0} and nonlinearity coefficient ratio c for two different κ_{i0}/κ_{h0} ratios (10 and 0.1). First, Fig. 6.3 shows the effective linear thermal conductivity κ_{e0} as a function versus f_i. In Fig. 6.3a, $\kappa_{i0}/\kappa_{h0} = 10$ while $\kappa_{i0}/\kappa_{h0} = 0.1$ in Fig. 6.3b. It can be seen that κ_{e0} is always between the values of κ_{i0} and κ_{h0}. Also, the Rayleigh

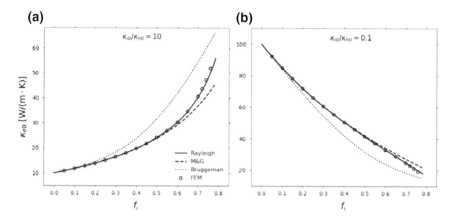

Fig. 6.3 The linear part of effective thermal conductivity κ_{e0} versus concentration f_i. In **a**, $\kappa_{i0} = 100$ W/(m K) and $\kappa_{h0} = 10$ W/(m K) while $\kappa_{i0} = 10$ W/(m K) and $\kappa_{h0} = 100$ W/(m K) in (**b**). The light blue, black and red lines represent the analytical predictions of the Rayleigh method, Maxwell-Garnett formula and Bruggeman formula respectively. The scatter plot of blue circles shows the finite-element simulation results. Adapted from Ref. [3]

method is more accurate than effective medium theories (both Maxwell-Garnett formula and Bruggeman formula), especially for big values of f_i. This conclusion results from the fact that the effective medium theories are derived based on the assumption that the inclusions are randomly distributed and the multipolar interactions (beyond dipolar interactions) between inclusions are neglected. However, the Rayleigh method considers such interactions within periodic structures by taking the Rayleigh identity as part of the boundary conditions. As we know, when f_i is not dilute, multipolar interactions between inclusions can't be neglected.

Figure 6.4a, b show the nonlinearity coefficient ratio c for nonlinear inclusions embedded in a linear host (the first case illustrated by Fig. 6.2a) while (c, d) show the results for linear inclusions embedded in a nonlinear host (the second case illustrated by Fig. 6.2b). The linear part of effective conductivity for Fig. 6.4a, c ($\kappa_{i0}/\kappa_{h0} = 10$) can be found in Fig. 6.3a and that for Fig. 6.4b, d ($\kappa_{i0}/\kappa_{h0} = 0.1$) can be found in Fig. 6.3b. Again, the Rayleigh method provides more accurate predictions than effective medium theories.

6.4 Exercises and Solutions

Exercises

1. Prove Eq. (6.8), meaning

$$\langle \nabla T_i \rangle = \frac{2\kappa_h}{\kappa_i + \kappa_h} \langle \nabla T_h \rangle . \tag{6.60}$$

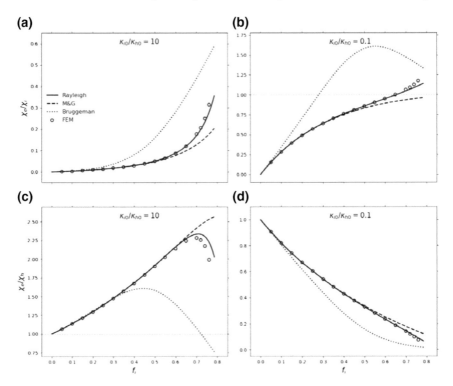

Fig. 6.4 Nonlinearity coefficient ratio versus f_i. **a** and **b** correspond to Fig. 6.2a where only the inclusions are nonlinear while (c) and (d) correspond to Fig. 6.2b where only the host is nonlinear. In **a**, $\kappa_h = 10$ W/(m K) and $\kappa_i = 100$ W/(m K) $+ \left[0.01 \text{ W/(m K}^2)\right] \times T$. In **b**, $\kappa_h = 100$ W/(m K) and $\kappa_i = 10$ W/(m K) $+ \left[0.001 \text{ W/(m K}^2)\right] \times T$. In **c**, $\kappa_h = 10$ W/(m K) $+ \left[0.001 \text{ W/(m K}^2)\right] \times T$ and $\kappa_i = 100$ W/(m K). In **d**, $\kappa_h = 100$ W/(m K) $+ \left[0.01 \text{ W/(m K}^2)\right] \times T$ and $\kappa_i = 10$ W/(m K). Adapted from Ref. [3]

2. Consider a two-dimensional structure and a thermal conduction distribution as

$$\begin{cases} \kappa(r) = ar^b, & 0 \le r \le 1 \\ \kappa(r) = 1, & r > 1 \end{cases} \tag{6.61}$$

   The region $0 \le r \le 1$ is the inclusion and the rest area denotes the host. Calculate κ_e in the x direction ($b > 0$).
3. What are the 2nd-order term of heat flux and governing equation for nonlinear heat conduction?

Solutions

1. **Solution**: Since the inclusions are dilute, we again obtain Eq. (6.49)

$$
\begin{cases}
T_h^{(0)} = G\dfrac{x}{x^2 + y^2} + Hx \\
T_i^{(0)} = Cx
\end{cases},
\tag{6.62}
$$

where

$$
\begin{cases}
G = Hr_i^2\dfrac{\kappa_{h0} - \kappa_{i0}}{\kappa_{h0} + \kappa_{i0}} \\
C = 2H\dfrac{\kappa_{h0}}{\kappa_{h0} + \kappa_{i0}}
\end{cases}.
\tag{6.63}
$$

Notice the term $G\dfrac{x}{x^2 + y^2}$ results from the interaction between inclusions and the host, which can be negligible compared with Hx so we have

$$
\nabla T_i = \frac{2\kappa_h}{\kappa_i + \kappa_h}\nabla T_h,
\tag{6.64}
$$

and thus

$$
\langle \nabla T_i \rangle = \frac{2\kappa_h}{\kappa_i + \kappa_h}\langle \nabla T_h \rangle.
\tag{6.65}
$$

2. **Solution**: In polar coordinates, temperature satisfies

$$
\frac{1}{r}\frac{\partial}{\partial r}\left(r\kappa(r)\frac{\partial T}{\partial r}\right) + \frac{\kappa(r)}{r^2}\frac{\partial^2 T}{\partial \theta^2} = 0.
\tag{6.66}
$$

Using separation of variables $(T = R(r)\Theta(\theta))$, we have

$$
r^2\left(\frac{d^2 R}{dr^2} + \frac{1}{\kappa(r)}\frac{d\kappa(r)}{dr}\frac{dR}{dr}\right) + r\frac{dR}{dr} - n^2 R = 0,
\tag{6.67}
$$

and

$$
\frac{d^2\Theta}{d\theta^2} + n^2\Theta = 0.
\tag{6.68}
$$

Taking the form of κ into Eq. (6.49) we have

$$
r^2\frac{d^2 R}{dr^2} + (b+1)r\frac{dR}{dr} - n^2 R = 0
\tag{6.69}
$$

and it's homogeneous. Then we can set $R(r) = r^s$ and get

$$
s_{\pm}(n) = \frac{1}{2}\left(-b \pm \sqrt{b^2 + 4n^2}\right).
\tag{6.70}
$$

According to boundary conditions ($T_{r\to\infty} = -J_0 r \cos(\theta)$), we have

$$T_i = T_0 + A_1 r^{s_+(1)} \cos(\theta) \tag{6.71}$$

and

$$T_m = T_0 - J_0 r \cos(\theta) + \frac{D_1}{r} \cos(\theta), \tag{6.72}$$

where

$$A_1 = -\frac{2}{sa+1} J_0, \quad D_1 = \frac{sa-1}{sa+1} J_0. \tag{6.73}$$

Finally we can obtain that

$$2\pi \frac{\kappa_e - 1}{\kappa_e + 1} J_0 = \int_S [\kappa(r) - 1] \frac{\partial T}{\partial x} dS \tag{6.74}$$

and

$$\kappa_e = \frac{a\left(s^2 + (b+2)s + 1\right)}{a\left(s^2 + bs - 1\right) + 2(s+b+1)}. \tag{6.75}$$

3. **Solution**: The 2nd-order nonlinear heat flux and conduction equation are respectively

$$\mathbf{J}^{(2)} = -\kappa_0 \nabla T^{(2)} - \kappa_1 T^{(0)} \nabla T^{(1)} - \kappa_1 T^{(1)} \nabla T^{(0)} - \kappa_2 (T^{(0)})^2 \nabla T^{(0)}, \tag{6.76}$$

and

$$\rho c \frac{\partial T^{(2)}}{\partial t} = \nabla \cdot (\kappa_0 \nabla T^{(2)} + \kappa_1 T^{(0)} \nabla T^{(1)} + \kappa_1 T^{(1)} \nabla T^{(0)} + \kappa_2 (T^{(0)})^2 \nabla T^{(0)}). \tag{6.77}$$

References

1. Garnett, J.C.M.: Colours in metal glasses and in metallic films. Philos. Trans. R. Soc. London Ser. A **203**, 385 (1904)
2. Bruggeman, D.A.G.: Berechnung verschiedener physikalischer Konstanten von heterogenen substanzen. I. Dielektrizitätskonstanten und Leitfähigkeiten der Mischkörper aus isotropen Substanzen (Calculation of different physical constants of heterogeneous substances. I. Dielectricity and conductivity of mixtures of isotropic substances). Annalen der Physik **24**, 636–664 (1935)
3. Dai, G.L., Huang, J.P.: Nonlinear thermal conductivity of periodic composites. Int. J. Heat Mass Transfer **147**, 118917 (2020)

4. Xu, L.J., Jiang, C.R., Shang, J., Wang, R.Z., Huang, J.P.: Periodic composites: quasi-uniform heat conduction, Janus thermal illusion, and illusion thermal diodes. Eur. Phys. J. B **90**, 221 (2017)
5. Rayleigh, L.: On the influence of obstacles arranged in rectangular order upon the properties of a medium. Philos. Mag. **34**, 481–502 (1892)
6. Gu, G.Q., Yu, K.W., Hui, P.M.: First-principles approach to conductivity of a nonlinear composite. Phys. Rev. B **58**, 3057 (1998)
7. Gu, G.Q., Yu, K.W.: Effective conductivity of nonlinear composites. Phys. Rev. B **46**, 4502 (1992)
8. Yu, K.W., Gu, G.Q.: Effective conductivity of nonlinear composites. II. Effective-medium approximation. Phys. Rev. B **47**, 7568 (1993)
9. Dai, G.L., Shang, J., Wang, R.Z., Huang, J.P.: Nonlinear thermotics: nonlinearity enhancement and harmonic generation in thermal metasurfaces. Eur. Phys. J. B **91**, 59 (2018)

Chapter 7
Heat Conduction Equation

Abstract Manipulating thermal conductivities are fundamentally important for controlling the conduction of heat at will. Thermal cloaks and concentrators, which have been extensively studied recently, are actually graded materials designed according to coordinate transformation approaches, and their effective thermal conductivity can be seen to equal that of the host medium outside the cloak or concentrator. Here we attempt to investigate a more general problem: what is the effective thermal conductivity of graded materials? In particular, we perform a first-principles approach to the analytic exact results of effective thermal conductivities of materials possessing either power-law or linear gradation profiles. On the other hand, by solving Laplace's equation, we derive a differential equation for calculating the effective thermal conductivity of a material whose thermal conductivity varies along the radius with arbitrary gradation profiles. The two methods agree well with each other for both external and internal heat sources, as confirmed by simulations and experiments. This chapter provides different methods for designing new thermal metamaterials (including thermal cloaks and concentrators), in order to control or manipulate the transfer of heat

Keywords Laplace's equation · Effective thermal conductivity · Graded materials · Gradation profiles

7.1 Opening Remarks

Thermal conductivity is the fundamental physical parameter that describes the ability of a material to conduct heat. How to design the distribution of thermal conductivities is particularly important for obtaining new kinds of thermal metamaterials [1–15] (the concept of metamaterial has been widely adopted as a material structurally designed to have a new property or function other than naturally occurring materials or chemical compounds), like thermal cloaks [1, 2, 5, 7, 9, 16, 17], thermal concentrators [5, 8], thermal transparency [6], macroscopic thermal diodes [10], and energy-free thermostat [11].

© Springer Nature Singapore Pte Ltd. 2020
J.-P. Huang, *Theoretical Thermotics*,
https://doi.org/10.1007/978-981-15-2301-4_7

However, according to the transformation theory of thermal conduction (which is based on the fact that the thermal conduction equation fulfills form invariance under coordinate transformations) [1], all the thermal cloaks [1, 2, 5, 7, 9] and thermal concentrators [5, 8] are essentially graded materials whose thermal conductivities vary along the radius. Moreover, their effective thermal conductivities equal to those of the host medium outside the cloak or concentrator. As a result, the existence of cloaks or concentrators does not affect the distribution of temperature or heat flux in the host medium, thus yielding a kind of thermal invisibility. This encourages us to ask a more general problem: what is the effective thermal conductivity of graded materials with arbitrary gradation profiles? This has not been touched in the literature due to the lack of suitable methods. In this chapter, we manage to solve this problem, in order to control or manipulate heat transfer with a different degree of freedom.

7.2   Analytic Theory Based on a First-Principles Approach

We consider a graded circular material with radius r_0 subjected to a uniform density of heat flux J_0 along the x-axis, the temperature distribution of the system satisfies the thermal conduction equation depending on time t, $\nabla \cdot \mathbf{J} + \rho c \frac{\partial T}{\partial t} = Q$. Here, J, T and Q represent the density of heat flux, temperature, and heat energy generated per unit volume per unit time, respectively. ρ denotes the mass density of the object and c is the specific heat capacity. Using the Fourier law, $J = -\kappa(r)\nabla T$ (where $\kappa(r)$ is the thermal conductivity of the material, which is a function of the position r along the radius, $r \leq r_0$), for static cases without internal heat sources, the above thermal conduction equation reduces into

$$\nabla \cdot [\kappa(r)\nabla T] = 0. \tag{7.1}$$

According to Eq. (7.1) in polar coordinates (r, θ), the temperature T satisfies

$$\frac{1}{r}\frac{\partial}{\partial r}\left(r\kappa(r)\frac{\partial T}{\partial r}\right) + \frac{\kappa(r)}{r^2}\frac{\partial^2 T}{\partial \theta^2} = 0. \tag{7.2}$$

If we write $T = R(r)\Theta(\theta)$ to achieve the separation of variables, we obtain

$$r^2\left(\frac{d^2 R}{dr^2} + \frac{1}{\kappa(r)}\frac{d\kappa(r)}{dr}\frac{dR}{dr}\right) + r\frac{dR}{dr} - n^2 R = 0, \tag{7.3}$$

and

$$\frac{d^2\Theta}{d\theta^2} + n^2\Theta = 0. \tag{7.4}$$

Without loss of generality, we set both r_0 and the thermal conductivity outside the material to be unit. If the thermal conductivity of the material has specific gradation profiles, the exact solution can be obtained by using the first-principles approach. For example, we give two examples in the following.

7.2.1 Exact Solution for Thermal Conductivity Distributed in a Power-Law Profile

Assume that the thermal conductivity of the material increases outwards in a power-law form. In this case, $\kappa(r) = ar^b$ (here a and b are two coefficients, $b \geq 0$; $0 \leq r \leq 1$), then Eq. (7.3) becomes

$$r^2 \frac{d^2 R}{dr^2} + (b+1)r \frac{dR}{dr} - n^2 R = 0. \tag{7.5}$$

Since this equation is homogeneous, the solution has the form as $R(r) = r^s$. Substituting it into Eq. (7.5) yields

$$s_{\pm}(n) = \frac{1}{2} \left(-b \pm \sqrt{b^2 + 4n^2} \right). \tag{7.6}$$

In the far field where the host medium has a thermal conductivity of $\kappa_m = 1$, the temperature is only determined by $\vec{J_0}$, which means $T_{r \to \infty} = -J_0 r \cos(\theta)$. In the material, the condition of convergence ensures that $T_{r \to 0} = $ finite value. So the terms for $s \geq 2$ vanish. The temperature fields in the material and host medium are respectively given by

$$T_i = T_0 + A_1 r^{s_+(1)} \cos(\theta), \tag{7.7}$$

and

$$T_m = T_0 - J_0 r \cos(\theta) + \frac{D_1}{r} \cos(\theta). \tag{7.8}$$

The coefficients are determined by the associated boundary conditions,

$$T_i|_{r=1} = T_m|_{r=1}, \quad \kappa(r) \frac{dT_i}{dr}\Big|_{r=1} = \frac{dT_m}{dr}\Big|_{r=1}. \tag{7.9}$$

As a result, we obtain

$$A_1 = -\frac{2}{sa+1} J_0, \quad D_1 = \frac{sa-1}{sa+1} J_0. \tag{7.10}$$

Since both the gradation profile and the temperature boundary condition are symmetric, we concern more about the space variation of the temperature field along the x-axis, which can be written as

$$\frac{\partial T}{\partial x} = -A_1 r^{s-1}((s-1)\cos^2(\theta)+1). \tag{7.11}$$

To analyze the response of the material to the external temperature field, we introduce the effective thermal conductivity κ_e. If the thermal conductivity distributed in the material is replaced by the uniform thermal conductivity κ_e, the value and gradient of the temperature at the boundary between the material and host medium will not change. In this case, the thermal medium with κ_e shows a dipolar effect on the external temperature field. So we obtain

$$2\pi \frac{\kappa_e - 1}{\kappa_e + 1} J_0 = \int_S [\kappa(r) - 1]\frac{\partial T}{\partial x} dS, \tag{7.12}$$

where S denotes the area occupied by the material. Calculating the above equation gives

$$\kappa_e = \frac{a\left(s^2 + (b+2)s + 1\right)}{a\left(s^2 + bs - 1\right) + 2(s+b+1)}. \tag{7.13}$$

If $b = 0$, $\kappa(r)$ is a constant, and $s = 1$. Then we achieve the desired result,

$$\kappa_e = a. \tag{7.14}$$

7.2.2 Exact Solution for Thermal Conductivity Distributed in a Linear Profile

We consider a linear gradation profile $\kappa(r) = cr + d$ for the graded material, where c and d are two coefficients. The analytic procedure is much the same as in Sect. 7.2.1. For the sake of simplicity, we set $\hat{r} = \frac{d}{c}r$. Then, the radial function follows

$$\frac{d^2 R}{d\hat{r}^2} + \left(\frac{1}{\hat{r}} + \frac{1}{\hat{r}+1}\right)\frac{dR}{d\hat{r}} - \frac{n^2 R}{\hat{r}^2} = 0. \tag{7.15}$$

The power series solution can be expressed as

$$f_n(\hat{r}) = \sum_{k=0}^{\infty} C_k^n \hat{r}^{k+\rho}. \tag{7.16}$$

Substituting it into Eq. (7.15) yields

$$\sum_{k=0}^{\infty} C_k^n \left[(k+\rho-1)(k+\rho) + (k+\rho) - n^2\right]\hat{r}^{k+\rho-2} -$$

$$\sum_{k=0}^{\infty} C_k^n \left[(k+\rho-1)(k+\rho)+2(k+\rho)-n^2\right] \hat{r}^{k+\rho-1} = 0.$$

The coefficient of each term should vanish. After solving the lowest term, we can easily get

$$\rho = \pm n, \tag{7.17}$$

and the recursion relation

$$C_{k+1}^n = -\frac{(k+n)(k+n+1)-n^2}{(k+n+1)^2-n^2} C_k^n. \tag{7.18}$$

The series should be convergent for seeking the exact solution. Therefore, we require the condition of linear profiles with a small slope, which means $\left|\frac{d}{c}\right| > 1$. Whereafter, the temperature fields in the material (T_i) and host medium (T_m) are respectively given by

$$T_i = T_0 + A_1 \sum_{k=0}^{\infty} C_k^1 \left(\frac{c}{d}\right)^{k+1} \cos(\theta), \tag{7.19}$$

$$T_m = T_0 + \frac{D_1 \cos(\theta)}{r} - J_0 r \cos(\theta), \tag{7.20}$$

where

$$A_1 = -\frac{2}{V_2(c+d)+V_1} J_0, \quad D_1 = \frac{V_2(c+d)-V_1}{V_2(c+d)+V_1} J_0. \tag{7.21}$$

Here

$$V_1 = \sum_{k=0}^{\infty} C_k^1 \left(\frac{c}{d}\right)^{k+1}, \quad V_2 = \sum_{k=0}^{\infty} (k+1) C_k^1 \left(\frac{c}{d}\right)^{k+1}. \tag{7.22}$$

On the other hand, solving the temperature field along the x-axis yields

$$\frac{\partial T}{\partial x} = -A_1 \sum_{k=0}^{\infty} C_k^1 \left(\frac{c}{d}\right)^{k+1} r^k \left(k \cos^2(\theta)+1\right). \tag{7.23}$$

The substitution of Eq. (7.23) into Eq. (7.12) yields the effective thermal conductivity

$$\kappa_e = \frac{c\,(V_2+V_3)+d\,(V_1+V_2)}{c\,(V_2-V_3)-(d-2)V_1+d V_2}, \tag{7.24}$$

where

$$V_3 = \sum_{k=0}^{\infty} \frac{(k+2)C_k^1 \left(\frac{c}{d}\right)^{k+1}}{k+3}. \tag{7.25}$$

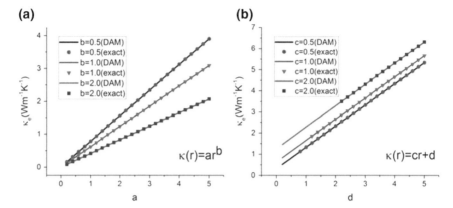

Fig. 7.1 Effective thermal conductivity κ_e for two gradation profiles: **a** $\kappa(r) = ar^b$ and **b** $\kappa(r) = cr + d$. **a** κ_e versus a for different b; **b** κ_e versus d for different c. The solid lines denote the results calculated from the DAM (Eq. 7.30); the symbols are exact results predicted from **a** Eq. (7.13) and **b** Eq. (7.24). Adapted from Ref. [18]

When $c = 0$, $\kappa(r)$ is a constant, and Eq. (7.24) reduces to the known case,

$$\kappa_e = d. \tag{7.26}$$

Now we are allowed to compare the exact solutions (Eqs. 7.13 and 7.24) with the results obtained from the differential equation (Eq. 7.30), in order to validate the above DAM. The numerical integration has been done by the fourth-order Runge-Kutta algorithm. Figure 7.1 shows power-law and linear gradation profiles of $\kappa(r)$ with various coefficients. Clearly the DAM (Eq. 7.30) agrees with the exact results predicted from Eq. (7.13) (Fig. 7.1a) and Eq. (7.24) (Fig. 7.1b), as expected. It is worth noting that the linear solutions should satisfy the small slope condition, which causes the lack of solutions when d is relatively small; see Fig. 7.1b.

7.3 Differential Approximation Method (DAM): A Differential Equation Approach

A graded material may be regarded differentially as a multi-layer structure. Let us start by considering a simple material that is composed of a homogeneous circular core (with thermal conductivity κ_c) plus a homogeneous circular shell (with κ_s). Solving Laplace's equation and the associated boundary conditions yields the following expression for its effective thermal conductivity κ_e,

$$\kappa_e = \kappa_s \frac{\kappa_c(1 + p) + \kappa_s(1 - p)}{\kappa_c(1 - p) + \kappa_s(1 + p)}, \tag{7.27}$$

where p is the area fraction of the core. For the sake of convenience, we re-write Eq. (7.27) as

$$\frac{\kappa_e - \kappa_s}{\kappa_e + \kappa_s} = p\frac{\kappa_c - \kappa_s}{\kappa_c + \kappa_s}. \tag{7.28}$$

On the other hand, we construct a graded material with radius r. Then, we encircle the material with a shell of infinitesimal thickness dr. The effective thermal conductivity changes from $\kappa_e(r)$ to $\kappa_e(r + dr)$. In this case, Eq. (7.28) helps to obtain

$$\frac{\kappa_e(r + dr) - \kappa(r)}{\kappa_e(r + dr) + \kappa(r)} = \frac{r^2}{(r + dr)^2}\frac{\kappa_e(r) - \kappa(r)}{\kappa_e(r) + \kappa(r)}. \tag{7.29}$$

Here $\kappa(r)$ is the thermal conductivity of the shell. Then, we obtain a differential equation,

$$\frac{d\kappa_e(r)}{dr} = \frac{\kappa(r)^2 - \kappa_e(r)^2}{r\kappa(r)}. \tag{7.30}$$

Given the gradation profile $\kappa(r)$ and the initial condition when radius is close to zero, the effective thermal conductivity of the whole graded circular material, $\kappa_e(r)$, can be achieved according to Eq. (7.30). This differential equation requires that the thermal conductivity of each shell cannot be zero, of which the first-order derivative should be continuous.

Incidentally, the differential equation for the effective thermal conductivity of a graded spherical material can be readily obtained on the same footing [19],

$$\frac{d\kappa_e(r)}{dr} = \frac{2\kappa(r)^2 - \kappa(r)\kappa_e(r) - \kappa_e(r)^2}{r\kappa(r)}. \tag{7.31}$$

7.4 Computer Simulations Based on a Finite-Element Method

By using COMSOL (https://www.comsol.com), we perform two-dimensional finite-element simulations to further confirm the validity of DAM. In the mean time, more detailed thermal responses of graded materials can be revealed. The basic parameters of our simulation system are set as follows. A graded circular material with the radius of 6 cm is embedded in the center of a square host medium with the side length of 20 cm. To maintain a uniform density of heat flux, the left side of the host medium holds a line hot source with temperature 313 K, while the right side 273 K.

Figure 7.2a, d, g show the simulation results for three different power-law gradation profiles. Figure 7.2b, e, h represent effective thermal materials of Fig. 7.2a, d, g respectively, whose thermal conductivities are computed according to both Eqs. (7.30) and (7.13) (the two equations give the same results). The thermal conductivity of the host medium in Fig. 7.2a, b, d, e, g, h has the same value, which

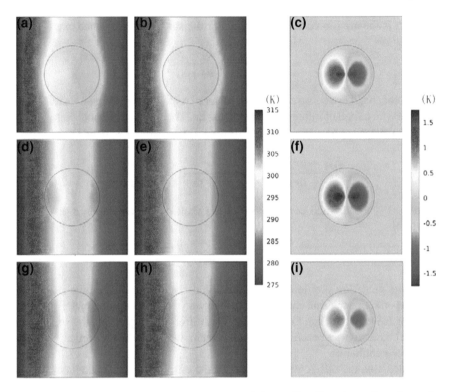

Fig. 7.2 Finite-element simulations for power-law gradation profiles. The color surfaces denote the distribution patterns of **a, b, d, e, g, h** temperature and **c, f, i** temperature difference, as represented by the associated color bar. The thermal conductivity of the materials is **a** $1.0r^2$, **b** $10.36 \, \mathrm{Wm^{-1}K^{-1}}$, **d** $1.0r^1$, **e** $3.09 \, \mathrm{Wm^{-1}K^{-1}}$, **g** $0.5r^1$, and **h** $1.74 \, \mathrm{Wm^{-1}K^{-1}}$; in **a, b, d, e, g, h**, the host medium has a thermal conductivity of $3.09 \, \mathrm{Wm^{-1}K^{-1}}$. **c, f**, and **i** show the temperature difference between **a** and **b**, **d** and **e**, and **g** and **h**, respectively. Adapted from Ref. [18]

equals the effective thermal conductivity of the graded material shown in Fig. 7.2d. Accordingly, we observe the different temperature patterns within the host medium areas in Fig. 7.2a, d, g or Fig. 7.2b, e, h. For more detailed comparison, Fig. 7.2c, f, i display the calculated difference between Fig. 7.2a and b, d and e, and g and h, respectively. Clearly, Fig. 7.2c, f, i show the zero value outside the circular material region, which further confirms the validity of (and agreement between) Eqs. (7.13) and (7.30).

The layout of Fig. 7.3 is roughly the same as Fig. 7.2, but for the graded material with linear gradation profiles in Fig. 7.3a, d, g. The thermal conductivities of materials in Fig. 7.3b, e, h are different, which respectively equal to the effective thermal conductivity of Fig. 7.3a, d, g according to Eq. (7.30) or Eq. (7.13). Similarly, Fig. 7.3c, f, i display the zero value outside the circle area, which also helps to validate Eqs. (7.30) and (7.13).

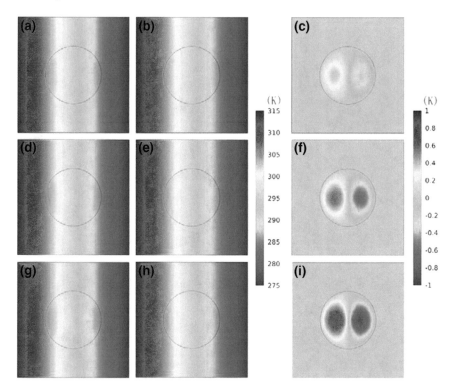

Fig. 7.3 Finite-element simulations for linear gradation profiles. The thermal conductivities of the host medium and the material are **a** 5.64 $Wm^{-1}K^{-1}$ and $0.5r + 4$, **b** 5.64 $Wm^{-1}K^{-1}$ and 5.64 $Wm^{-1}K^{-1}$, **d** 7.23 $Wm^{-1}K^{-1}$ and $1.0r + 4$, **e** 7.23 $Wm^{-1}K^{-1}$ and 7.23 $Wm^{-1}K^{-1}$, **g** 10.39 $Wm^{-1}K^{-1}$ and $2.0r + 4$, and **h** 10.39 $Wm^{-1}K^{-1}$ and 10.39 $Wm^{-1}K^{-1}$. **c, f** and **i** display the temperature difference between **a** and **b**, **d** and **e**, and **g** and **h**, respectively. Adapted from Ref. [18]

So far, both Figs. 7.2 and 7.3 have shown that the DAM (Eq. 7.30) works well under the conditions of power-law or linear gradation profiles of thermal conductivity. Actually, the DAM is applicable for arbitrary gradation profiles, including multi-layer structures. See Fig. 7.4. Figure 7.4 has the same layout as Fig. 7.3, but displaying three multi-layer structures in Fig. 7.4a, d, g. The thermal conductivities adopted for the circular materials in Fig. 7.4b, e, h are educed by the DAM (Eq. 7.30) for the multi-layer structures displayed in Fig. 7.4a, d, g respectively. Note that by choosing the layer thicknesses appropriately, the effective thermal conductivities of the three multi-layer structures are exactly the same in Fig. 7.4(a,d,g), as calculated by Eq. (7.30). Clearly, Fig. 7.4c, f, i also display the zero value outside the multi-layer structure, which validates Eq. (7.30) for the multi-layer structure indeed.

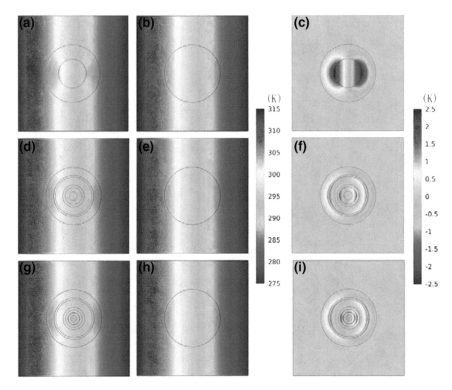

Fig. 7.4 Finite-element simulations for multi-layer profiles. In **a**, **d**, **g**, the multi-layer material is made of two materials (with thermal conductivity 10 and 90 Wm^{-1}K^{-1}) in alternation: **a** two layers, **d** six layers, and **g** ten layers; the central layer of **a**, **d**, **g** has the thermal conductivity of 10 Wm^{-1}K^{-1}. In **a**, **b**, **d**, **e**, **g**, **h**, the thermal conductivity of the host medium is 60 Wm^{-1}K^{-1}. **c**, **f** and **i** represent the temperature difference between **a** and **b**, **d** and **e**, and **g** and **h**, respectively. Adapted from Ref. [18]

7.5   Experiments Based on a Multi-layer Circular Structure

In order to further confirm the validity of the DAM (Eq. 7.30), here we experimentally investigate a multi-layer material. Our experimental design is shown in Fig. 7.5a, d. Figure 7.5a contains a six-layer material, which is made of two materials (copper and phosphor bronze) in alternation. For comparison, Fig. 7.5d includes a homogeneous material (brass) with the thermal conductivity (109 Wm^{-1}K^{-1}) equal to the effective thermal conductivity of the multi-layer material shown in Fig. 7.5a calculated according to Eq. (7.30). The left-hand side of the host medium (copper) is connected with hot water, and the right-hand side immersed into cold water. A thermal imager is emplaced right above the multi-layer material. The experiment is conducted in the air. Air convection and thermal contact resistance are two possible factors affecting experimental accuracy, which, however, can be reduced by using appropriate approaches (e.g., fine welding). Figure 7.5b, e show the finite-element simulations of

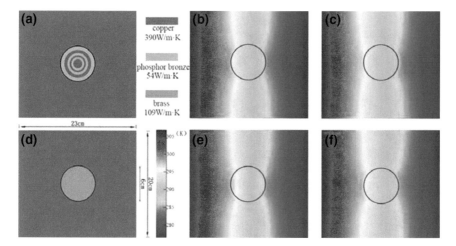

Fig. 7.5 Experimental results of a multi-layer material. **a** Experimental structure, **b** finite-element simulation of **a**, and **c** experimental measurement results of **a**. **d**, **e** and **f** are the reference group of **a**, **b** and **c**, respectively. The thickness of the experimental structures shown in **a**, **d** is 0.03 cm; other parameters are indicated in the figure. Adapted from Ref. [18]

Fig. 7.5a, d, respectively. Figure 7.5c, f exhibit the experimental results of Fig. 7.5a, d, respectively. Clearly the experimental results (Fig. 7.5c, f) echo with the simulation results (Fig. 7.5b, e). Importantly, the temperature distribution patterns in Fig. 7.5b, c are similar to those in Fig. 7.5e, f. This behavior indicates that our experimental results support the DAM (Eq. 7.30) indeed.

7.6 Discussion and Conclusions

We have derived both a first-principles approach and a DAM (differential approximation method; Eq. 7.30) for calculating the effective thermal conductivity of a circular material whose thermal conductivity varies along the radius with specific or arbitrary gradation profiles. This Equation (7.30) has been confirmed by analytic theory (based on a first-principles approach), computer simulations (based on a finite-element method), and experiments (based on a multi-layer circular structure).

Self-heating objects are common in nature, such as human bodies or electric equipments. Our DAM Eq. (7.30) may hold for such self-heating cases under some conditions. For example, let us introduce a kind of self-heating multi-layer material and deduce the effective thermal conductivity. Here the self-heating means that the center of the multi-layer material is keeping at a constant temperature, which can be seen as another boundary condition in the thermal model. Meanwhile, the multi-layer material is located in a uniform density of heat flux along x-axis. What we aspire herein is that the thermal response of the self-heating multi-layer material is just the

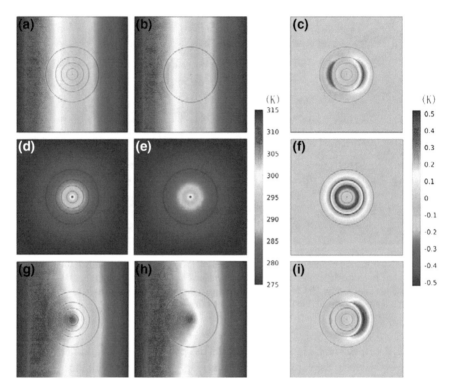

Fig. 7.6 Finite-element simulations for self-heating multi-layer materials. In **a**, **b**, **d**, **e**, **g**, **h**, the thermal conductivity of the host medium is $300\ \mathrm{Wm^{-1}K^{-1}}$. In **a**, **d**, **g**, the 4-layer structure is made of materials with thermal conductivity 300, 275, 390 and 275 $\mathrm{Wm^{-1}K^{-1}}$ from inside-out. **a** A 4-layer material (without self-heating) subjected to a uniform density of heat flux; **d** a self-heating 4-layer material; **g** a self-heating 4-layer material subjected to a uniform density of heat flux. In **b**, the homogeneous circle's thermal conductivity ($300\ \mathrm{Wm^{-1}K^{-1}}$) is equal to the effective thermal conductivity of the multi-layer material shown in **a** determined by Eq. (7.30); in **e**, the homogeneous circle's thermal conductivity ($300\ \mathrm{Wm^{-1}K^{-1}}$) equals the effective thermal conductivity of the multi-layer material shown in **d** determined by Eq. (7.32); in **h**, the homogeneous circle's thermal conductivity ($300\ \mathrm{Wm^{-1}K^{-1}}$) equals the effective thermal conductivity of the multi-layer material shown in **g** determined by either Eq. (7.30) or Eq. (7.32). **c**, **f** and **i** display the temperature difference between **a** and **b**, **d** and **e**, and **g** and **h**, respectively. Adapted from Ref. [18]

same as a homogeneous material. On one hand, when there is no self-heating, we may resort to the DAM Eq. (7.30). The corresponding simulation results are shown in Fig. 7.6a–c. On the other hand, we need to make sure that the self-heating multi-layer material can be replaced by a homogeneous material. Considering the boundary conditions, we may derive the effective thermal conductivity κ_e at the view of the center of material as

$$\kappa_e = \ln\frac{r_n}{r_1}\left(\sum_{i=1}^{n-1}\frac{1}{\kappa_i}\ln\frac{r_{i+1}}{r_i}\right)^{-1},\qquad(7.32)$$

in which n is the total number of layers, and i is the serial number of each layer (with radius r_i and conductivity κ_i) of the multi-layer material from inside-out. Figure 7.6d–f show the simulation results. If the effective thermal conductivities calculated from the above two approaches (namely, Eqs. 7.30 and 7.32) are coincidently identical, we can safely superpose the thermal effects induced by these two kinds of heat sources adopted in Fig. 7.6a, d. As a result, Fig. 7.6g depicts a self-heating multi-layer material subjected to a uniform density of heat flux, which behaves just like a homogeneous material as shown in Fig. 7.6h, i.

This chapter is useful for designing new thermal metamaterials (including or going beyond thermal cloaks and thermal concentrators) for controlling/manipulating heat transfer, say, yielding the behavior of thermal transparency [6] when thermal conductivities depend on temperature or not [20]. Also, it is helpful for interdisciplinary researches on other kinds of gradation profiles when Laplace's equation governs the system [21–23].

7.7 Exercises and Solutions

Exercises

1. Prove Eq. (7.31).

Solutions

1. **Solution:** Refer to the article "L. Dong, J. P. Huang, K. W. Yu, and G. Q. Gu, Multipole polarizability of a graded spherical particle, Eur. Phys. J. B **48**, 439–444 (2005)", where the Eqs. (13)–(16) just show the desired solution by taking the $l = 1$ limit.

References

1. Fan, C.Z., Gao, Y., Huang, J.P.: Shaped graded materials with an apparent negative thermal conductivity. Appl. Phys. Lett. **92**, 251907 (2008)
2. Chen, T.Y., Weng, C.N., Chen, J.S.: Cloak for curvilinearly anisotropic media in conduction. Appl. Phys. Lett. **93**, 114103 (2008)
3. Li, J.Y., Gao, Y., Huang, J.P.: A bifunctional cloak using transformation media. J. Appl. Phys. **108**, 074504 (2010)
4. Guenneau, S., Amra, C., Veynante, D.: Transformation thermodynamics: cloaking and concentrating heat flux. Opt. Express **20**, 8207–8218 (2012)
5. Narayana, S., Sato, Y.: Heat flux manipulation with engineered thermal materials. Phys. Rev. Lett. **108**, 214303 (2012)
6. He, X., Wu, L.Z.: Thermal transparency with the concept of neutral inclusion. Phys. Rev. E **88**, 033201 (2013)
7. Schittny, R., Kadic, M., Guenneau, S., Wegener, M.: Experiments on transformation thermodynamics: molding the flow of heat. Phys. Rev. Lett. **110**, 195901 (2013)

8. Han, T.C., Zhao, J.J., Yuan, T., Lei, D.Y., Li, B.W., Qiu, C.W.: Theoretical realization of an ultra-efficient thermalenergy harvesting cell made of natural materials. Energ. Environ. Sci. **6**, 3537–3541 (2013)
9. Han, T.C., Bai, X., Gao, D.L., Thong, J.T.L., Li, B.W., Qiu, C.-W.: Experimental demonstration of a bilayer thermal cloak. Phys. Rev. Lett. **112**, 054302 (2014)
10. Li, Y., Shen, X.Y., Wu, Z.H., Huang, J.Y., Chen, Y.X., Ni, Y.S., Huang, J.P.: Temperature-dependent transformation thermotics: from switchable thermal cloaks to macroscopic thermal diodes. Phys. Rev. Lett. **115**, 195503 (2015)
11. Shen, X.Y., Li, Y., Jiang, C.R., Huang, J.P.: Temperature trapping: energy-free maintenance of constant temperatures as ambient temperature gradients change. Phys. Rev. Lett. **117**, 055501 (2016)
12. Shen, X.Y., Chen, Y.X., Huang, J.P.: Thermal magnifier and minifier. Commun. Theor. Phys. **65**, 375–380 (2016)
13. Xiang, Y.J., Wen, S.C., Dai, X.Y., Fan, D.Y.: Modulation instability in nonlinear oppositely directed coupler with a negative-index metamaterial channel. Phys. Rev. E **82**, 056605 (2010)
14. Acreman, A., Kaczmarek, M., D'Alessandro, G.: Gold nanoparticle liquid crystal composites as a tunable nonlinear medium. Phys. Rev. E **90**, 012504 (2014)
15. Reyes-Gomez, E., Cavalcanti, S.B., Oliveira, L.E.: Non-Bragg-gap solitons in one-dimensional Kerr-metamaterial Fibonacci heterostructures. Phys. Rev. E **91**, 063205 (2015)
16. Wu, L.Z.: Cylindrical thermal cloak based on the path design of heat flux. J. Heat Transfer **137**, 021301 (2015)
17. Xu, G.Q., Zhang, H.C., Zou, Q., Jin, Y.: Predicting and analyzing interaction of the thermal cloaking performance through response surface method. Int. J. Heat Mass transfer **109**, 746–754 (2017)
18. Ji, Q., Huang, J.P.: Controlling thermal conduction by graded materials. Commun. Theor. Phys. **69**, 434–440 (2018)
19. Milton, G.W.: The Theory of Composites. Cambridge University Press, Cambridge (2002)
20. Li, Y.Y., Li, N.B., Li, B.W.: Temperature dependence of thermal conductivities of coupled rotator lattice and the momentum diffusion in standard map. Eur. Phys. J. B **88**, 182 (2015)
21. Dong, L., Gu, G.Q., Yu, K.W.: First-principles approach to dielectric response of graded spherical particles. Phys. Rev. B **67**, 224205 (2003)
22. Dong, L., Huang, J.P., Yu, K.W., Gu, G.Q.: Multipole polarizability of a graded spherical particle. Eur. Phys. J. B **48**, 439 (2005)
23. Fan, C.Z., Gao, Y.H., Gao, Y., Huang, J.P.: Apparently negative electric polarization in shaped graded dielectric metamaterials. Commun. Theor. Phys. **53**, 913–919 (2010)

Chapter 8
Thermal Band Theory

Abstract In non-metal materials, phonon transport is the main mechanism of heat transfer and contributes to the most of thermal conductivity. This chapter describes the heat conduction from microscopic and mesoscopic levels, and presents some basic ideas of calculating thermal conductivities from the Boltzmann equation and the concept of engineering phonon band structure. Finally, thermocrystals are introduced, which have potential applications in guiding heat flux.

Keywords Thermal conductivity · Phonon transport · Phononic crystal ·
The Boltzmann equation · Band gap · Thermocrystals

8.1 Opening Remarks

This chapter mainly introduces the concept of thermocrystal proposed by Maldovan in Ref. [1]. It used a familiar periodic-composites structure of phononic crystal or sonic crystal while its lattice constant is at 10 nm level, which is much smaller than sound wavelength. Such a thermocrystal might be used to control heat in the manner of controlling wave.

Artificial periodic structures have been widely used to manipulate light and sound wave, which are known as photonic and phononic crystals respectively, as Bragg scattering can generate a photonic or phononic band gap to forbidden the transport of photons or phonons at some certain frequencies. A question is that can we also use artificial periodic structures called thermocrystals to control heat flux?

First we should make clear the carriers of heat. Phonons, which are the quantization of collective vibration of the lattices, contribute to most of the thermal conductivity in non-metallic materials. However, is a phononic crystal equivalent to a thermocrystal? What is the difference between phonons carrying heat and sound waves?

It can be noticed that sound waves are formed by the vibration of continuous media while phonons come from lattice vibration, a discrete model. In fact, only when wave vector or frequency is very small, lattice vibration can be approximately seen as a continuum mechanics model. Also, sound waves usually have low frequencies while in common materials heat is carried mainly by high-frequency phonons (more

© Springer Nature Singapore Pte Ltd. 2020   83
J.-P. Huang, *Theoretical Thermotics*,
https://doi.org/10.1007/978-981-15-2301-4_8

than THz at room temperature). The corresponding wavelengths of thermal phonons are usually about 2 nm and it is hard for current nanotechnology to manufacture a phononic crystal with such a small periodicity.

In addition, high-frequency phonons behave more like particles and can be scattered more easily than low-frequency phonons. As we can see in paragraphs below, thermal conductivity should be infinite without scattering because mean free paths are infinite in that case. In fact, scattering can be seen as anharmonic interaction between atoms and it is usually neglected in sound wave equation. So if we want to use a periodic phononic crystal to manipulate thermal phonons just like manipulating waves, we should enhance the proportion of low-frequency phonons in heat transfer and then take the lattice as continuous media. This is what Maldovan did in his article on thermocrystal [1]. To get a narrow phonon spectrum for heat transfer in bulk material silicon, he introduced impurities (nanoparticles) to shorten the mean free paths (which is proportional to thermal conductivity) of high-frequency phonons and enhance boundary scattering to shorten the mean free paths of extra low-frequency phonons. In this way, low-frequency phonons at about 100–300 GHz contribute most of the thermal conductivity and can be approximately taken as elastic waves. By arranging periodic holes in the bulk Si–Ge material, a phononic crystal with a band gap at about 180–300 GHz can be fabricated.

8.2 Boltzmann Transport Equation

To understand how phonons can affect heat transport, we will firstly talk about how to calculate thermal conductivities from the Boltzmann transport equation of phonons. The Boltzmann transport equation offers a method for calculating thermal conductivity from a mesoscopic level. Let us consider a three-dimensional crystal with N primitive cells and there are n atoms in each cell. Firstly, there exist 3 acoustic phonon modes or branches. The number 3 means 3 kinds of polarization: one longitudinal mode and two transverse modes. Besides, there are $3n - 3$ optical phonon modes if each primitive cell has n different atoms (sometimes the same kind of atoms in a cell can also generate optical modes). For each mode, wave vector $\boldsymbol{k}$ can have N discrete values in the first Brillouin zone when the total crystal has N primitive cells. For an ideal crystal with infinite primitive cells, $\boldsymbol{k}$ can be regarded as a continuous value and we can write the summation over all phonons with different $\boldsymbol{k}$ and polarization j as an integral form:

$$\frac{1}{V} \sum_{j,\boldsymbol{k}} \rightarrow \frac{1}{(2\pi)^3} \sum_{j} \int_{\mathrm{BZ}} \mathrm{d}\boldsymbol{k}, \tag{8.1}$$

where V is the volume of the whole lattice and "BZ" denotes the first Brillouin zone.

Let indices $j = 1, 2, 3$ represent acoustic phonon polarization, and then the Boltzmann transport equation for phonons distribution $f_{\boldsymbol{k},j}(\boldsymbol{r}, t)$ reads

$$\frac{\partial f_{k,j}(r,t)}{\partial t} + v_j(k) \cdot \nabla f_{k,j}(r,t) = \left(\frac{\partial f_{k,j}(r,t)}{\partial t}\right)_{scatt}. \tag{8.2}$$

Here t denotes time, $v_j(k)$ is the group velocity that equals $\dfrac{\partial \omega_j(k)}{\partial k}$, and $\omega_j(k)$ is the frequency of phonon for the j-th mode or branch at a given k. In addition, $\left(\frac{\partial f_{k,j}(r,t)}{\partial t}\right)_{scatt}$ is the scattering or collision term.

With relaxation-time approximation (the Callaway-Holland model), the scattering term has a simplified form as

$$\left(\frac{\partial f_{k,j}(r,t)}{\partial t}\right)_{scatt} = -\frac{f_{k,j}(r,t) - f_{k,j}^{eq}(r)}{\tau_j(k)}, \tag{8.3}$$

where $f_{k,j}^{eq}(r)$ is the equilibrium distribution or Bose-Einstein distribution of phonons

$$f_{k,j}^{eq}(r) = \frac{1}{\exp\left(\hbar\omega_j(k)/k_B T(r)\right) - 1}. \tag{8.4}$$

In addition, we assume $f_{k,j}(r)$ is near $f_{k,j}^{eq}(r)$ in steady states, meaning

$$f_{k,j}(r) = f_{k,j}^{eq}(r) + \delta f_{k,j}^{eq}(r), \tag{8.5}$$

and thus

$$\nabla f_{k,j}(r) \simeq \nabla f_{k,j}^{eq}(r) = \frac{\partial f_{k,j}^{eq}}{\partial T}\nabla T. \tag{8.6}$$

So we can rewrite

$$f_{k,j} \simeq f_{k,j}^{eq} - \tau_j(k)\frac{\partial f_{k,j}^{eq}}{\partial T}v_j(k) \cdot \nabla T. \tag{8.7}$$

The heat current density vector is

$$q = \frac{1}{V}\sum_{k,j} \hbar\omega_j(k)v_j(k)f_{k,j}, \tag{8.8}$$

or in the integral form as

$$q = \frac{1}{(2\pi)^3}\sum_{j}\int \hbar\omega_j(k)v_j(k)f_{k,j}dk. \tag{8.9}$$

Because heat flux is zero in equilibrium

$$\sum_{k,j} f_{k,j}^{eq}\hbar\omega_j(k)v_j(k) = 0, \tag{8.10}$$

we can rewrite heat flux density as

$$q = -\frac{1}{(2\pi)^3} \sum_j \int \hbar\omega_j(k) v_j(k)\tau_j(k) \frac{\partial f_{k,j}^{eq}}{\partial T} v_j(k) \cdot \nabla T \, dk$$

$$= -\frac{1}{(2\pi)^3} \sum_j \int v_j(k)\tau_j(k) v_j(k) \cdot \nabla T \frac{(\hbar\omega_j(k))^2 \exp(\hbar\omega_j(k)/k_B T)}{k_B T^2 \left[\exp(\hbar\omega_j(k)/k_B T) - 1\right]^2} \, dk.$$

$$(8.11)$$

Since the Fourier law of heat conduction writes

$$q = -\boldsymbol{\lambda} \cdot \nabla T, \tag{8.12}$$

the thermal conductivity tensor is

$$\boldsymbol{\lambda}(T) = \frac{1}{(2\pi)^3} \sum_j \int v_j(k) \otimes v_j(k)\tau_j(k) \frac{(\hbar\omega_j(k))^2 \exp(\hbar\omega_j(k)/k_B T)}{k_B T^2 \left[\exp(\hbar\omega_j(k)/k_B T) - 1\right]^2} \, dk \tag{8.13}$$

or

$$\lambda_{\alpha\beta}(T) = \frac{1}{(2\pi)^3} \sum_j \int (v_j(k))_\alpha (v_j(k))_\beta \tau_j(k) \frac{(\hbar\omega_j(k))^2 \exp(\hbar\omega_j(k)/k_B T)}{k_B T^2 \left[\exp(\hbar\omega_j(k)/k_B T) - 1\right]^2} \, dk. \tag{8.14}$$

Note that the phonon specific heat can be expressed as

$$C_{ph}(\omega_j(k)) \triangleq \frac{\partial}{\partial T} \left[\hbar\omega_j(k)(f_{k,j}^{eq} + \frac{1}{2}) \right]$$

$$= \frac{(\hbar\omega_j(k))^2 \exp(\hbar\omega_j(k)/k_B T)}{k_B T^2 \left[\exp(\hbar\omega_j(k)/k_B T) - 1\right]^2}. \tag{8.15}$$

And we add a useful approximation that the thermal conductivity is isotropic,

$$(v_j(k))_\alpha (v_j(k))_\beta \simeq v_j(k)^2 \langle \cos^2 \theta \rangle = \frac{1}{3}(v_j(k))^2, \tag{8.16}$$

where θ is the angle between $v_j(k)$ and temperature gradient. Then, the final form of the Callaway-Holland model in the wave vector space can be written as

$$\lambda(T) = \frac{1}{3}\frac{1}{(2\pi)^3} \sum_j \int (v_j(k))^2 \tau_j(k) C_{ph}(\omega_j(k)) \, dk$$

$$= \frac{1}{3}\frac{1}{(2\pi)^3} \sum_j \int v_j(k)\ell_j(k) C_{ph}(\omega_j(k)) \, dk. \tag{8.17}$$

Here $\ell_j(k) = v_j(k)\tau_j(k)$ is the mean free path. We can see that for a phonon with wave vector k and polarization j (and thus the frequency is determined), its contribution to thermal conductivity is proportional to the group speed, mean free path and phonon specific heat. As a result, if we want to block the phonon transport at a certain frequency range for reducing its contribution to the thermal conductivity, a convenient method is to shorten its mean free path by using enhanced scattering.

8.3 Scattering

If there is no scattering or collision, thermal conductivities would be infinite as relaxation time $\tau = \infty$ (here we neglect the indices k, j). Usually, the total relaxation time given by Matthiessen's rule is

$$\frac{1}{\tau} = \frac{1}{\tau_U} + \frac{1}{\tau_M} + \frac{1}{\tau_B} + \frac{1}{\tau_{ph-e}} + \cdots \tag{8.18}$$

Here τ_U, τ_M, τ_B, τ_{ph-e} is the relaxation time caused by Umklapp scattering, mass-difference impurity scattering,boundary scattering, phonon-electricity scattering respectively. Umklapp scattering is the anharmonic phonon interaction (usually consider third-order interaction or 3-phonons process only) which allows the violation of the conservation of phonon momentum. Umklapp scattering is stronger in high temperature and for high-frequency phonons, and usually we can write

$$\frac{1}{\tau_U} = ATe^{B/T}\omega^2, \tag{8.19}$$

where A and B are two coefficients. Mass-difference impurity scattering results from the impurities and lattice dislocation and can have a linear relationship with higher-order of phonon frequency:

$$\frac{1}{\tau_M} = D\omega^4. \tag{8.20}$$

Here D is a coefficient determined by the density difference of the matrix and impurity embedded, impurity particles radius and group speed. Boundary scattering is the result of irregular grain boundaries. Phonon-electricity scattering is quite complicated and often neglected for simplicity in studies. Matthiessen's rule also implies that these scattering processes are independent of each other.

However, sometimes Matthiessen's rule might not be so accurate especially for boundary scattering. Umklapp scattering, mass-difference impurity scattering and phonon-electricity scattering happen in the bulk while boundary scattering is a surface effect and relatively independent of the bulk effect. To improve the accuracy of prediction, Maldovan defined the (reduced) mean free path as

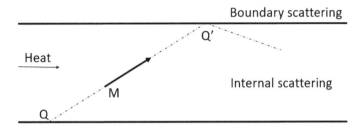

Fig. 8.1 Schematic graph showing boundary scattering and internal scattering in a thin film

$$\ell_j(\boldsymbol{k}) = \ell_{0j}(\boldsymbol{k}) \left[1 - \frac{[1 - p(\boldsymbol{k})] \exp\left[-L_1(\boldsymbol{k})/\ell_{0j}(\boldsymbol{k})\right]}{1 - p(\boldsymbol{k}) \exp\left[-L_2(\boldsymbol{k})/\ell_{0j}(\boldsymbol{k})\right]} \right], \qquad (8.21)$$

where $\ell_{0j}(\boldsymbol{k})$ is the bulk mean free path when boundary scattering is not considered, $p(\boldsymbol{k})$ is the boundary specularity and $L_1(\boldsymbol{k})$, $L_2(\boldsymbol{k})$ are characteristic lengths of the material. In a thin film, we can take $L_1(\boldsymbol{k}) = MQ'$ and $L_2(\boldsymbol{k}) = QQ'$ as shown in Fig. 8.1 [2].

8.4 Narrow Thermal Phonon Spectrum

Remember that our purpose is to construct a thermocrystal that can manipulate thermal phonons in the manner as manipulating waves like electromagnetic waves in photonic crystals or elastic waves in phononic crystals. To enhance the wave properties of phonons carrying heat, the corresponding frequencies should be low. According to Eq. (8.20), mass-difference impurity scattering is stronger for high-frequency phonons so Maldovan introduced Ge nanoparticles (with small diameters at nanometer level) in Ref. [1], which are embedded in bulk Si material. The filling fraction is 10% and thus a nano-structured alloy called $Si_{90}Ge_{10}$ is fabricated.

Next, Maldovan uses boundary scattering to selectively block extra low-frequency phonons and further narrow the thermal phonon spectrum. If the thickness of $Si_{90}Ge_{10}$ thin film is larger than medium-frequency and high-frequency phonons, only ultra-low frequency phonons will be significantly subjective to boundary scattering effect. Maldovan calculated the cumulative thermal conductivity for different thicknesses including 10, 100 and 1000 μm. The simulation results are shown in Fig. 1 of Ref. [1]. We can see that, for $Si_{90}Ge_{10}$ with nanoparticles (np), high-frequency phonons ($f > 1$ THz) contribute only about 15% of the conductivity while in the bulk Si they contribute about 80%. Also, the effect of blocking high-frequency phonons are better in nano-structured $Si_{90}Ge_{10}$ than in normal $Si_{90}Ge_{10}$. In addition, when the film gets thicker, extra-low frequency phonons carry less heat, yielding a narrower thermal phonon spectrum. When the thickness is 1000 μm, thermal phonons with frequency from 100 to 300 GHz carry 60% of the heat. In such a narrow hypersonic range, the mean free path of phonons is about 200–700 μm, which means the phonons can propagate relatively large distances and thus behave more like waves.

8.5 Thermal Band Gap

Now we have obtained a nano-structured $Si_{90}Ge_{10}$ alloy in which low-frequency phonons carrying heat can be regarded as elastic waves. The last step to design a thermocrystal is to use periodic composites just like phononic crystals. Firstly, since the phonon spectrum is about 100–300 GHz, we are allowed to write the elastic (sound) wave equation in an isotropic medium,

$$\rho \frac{\partial^2 u_i}{\partial t^2} = \nabla \cdot \left(\rho c_t^2 \nabla u_i \right) + \nabla \cdot \left(\rho c_t^2 \frac{\partial u}{\partial x_i} \right) + \frac{\partial}{\partial x_i} \left[\left(\rho c_l^2 - 2\rho c_t^2 \right) \nabla \cdot u \right], \quad (8.22)$$

where ρ is the density, $u = \sum u_i i$ is the displacement, and $c_t (or c_l)$ is the transverse (or longitudinal) speed of sound. In addition, with Poisson's ratio σ and Young's modulus E, we have

$$c_l = \sqrt{\frac{E(1 - \sigma)}{\rho(1 + \sigma)(1 - -2\sigma)}} \quad (8.23)$$

and

$$c_t = \sqrt{\frac{E}{2\rho(1 + \sigma)}}. \quad (8.24)$$

In such an (three-dimensional) isotropic medium, there are 3 acoustic modes, one longitudinal wave and two transverse waves, and each has a simple linear dispersion relationship. To obtain a phononic band gap, a commonly-used technique is placing periodic air holes in the medium and the periodicity should be near the wavelength of sound. By this way, the engineered material has more phonon modes. In Ref. [1], the lattice constant is set to be 10 and 20 nm and the air holes can be circular or other optimized shapes. From Fig. 2 of Ref. [1] that is the result of FEM simulation, we can see the band gap is just located in the narrow hypersonic frequency range of thermal phonons.

Based on such a thermocrystal, there are some potential applications including heat wave-guides and thermal diodes as illustrated in Fig. 3 of Ref. [1]. However, here we must point out that the band structure in Fig. 2 of Ref. [1] is not strict because we take the lattice wave as elastic wave as an approximation. This approximation can be used when phonons carrying heat are restricted to low-frequency range and the assumption of continuum mechanics is still valid. Luckily, a recent experiment tailoring thermal conductance with phononic crystal below 1 K shows that the measurement data are consistent with finite-element analysis of continuum mechanics [3].

8.6   Exercises and Solutions

Exercises

1. Phonon density of states (DOS) or vibration DOS for a certain frequency $g(\omega)$ is defined as

$$g(\omega) = \lim_{\Delta\omega \to 0} \frac{1}{\Delta\omega} \sum_{j,k}^{\omega \leqslant \omega_j(k) \leqslant \omega + \Delta\omega} = \sum_j \sum_k \delta\left(\omega - \omega_j(k)\right) \qquad (8.25)$$

or

$$g(\omega) = \lim_{\Delta\omega \to 0} \frac{1}{\Delta\omega} \frac{V}{(2\pi)^3} \sum_j \int_{\omega \leqslant \omega_j(k) \leqslant \omega + \Delta\omega} dk, \qquad (8.26)$$

which satisfies

$$\int g(\omega) d\omega = 3nN. \qquad (8.27)$$

Also we can define the DOS $g_j(\omega)$ for a certain mode or branch as

$$g_j(\omega) = \lim_{\Delta\omega \to 0} \frac{1}{\Delta\omega} \sum_{k}^{\omega \leqslant \omega_j(k) \leqslant \omega + \Delta\omega} = \sum_k \delta\left(\omega - \omega_j(k)\right). \qquad (8.28)$$

Here we take $g_j(\omega)$ and $g(\omega_j)$ as the same. Prove that the Callaway-Holland model in frequency space tells

$$\lambda(T) = \frac{1}{12\pi^4 V} \sum_j \int g(\omega_j) v_j^2(\omega_j) \tau_j(\omega_j) C_{\text{ph}}(\omega_j) d\omega. \qquad (8.29)$$

2. DOS of phonon is usually obtained by numerically simulation as the dispersion relationship can be quite complicated. Debye model is a useful approximation for low temperature limit, which takes the discrete lattice as continuous medium and use a linear dispersion relationship for acoustic modes

$$\omega_l = c_l k, \quad \omega_t = c_t k. \qquad (8.30)$$

Prove that: (i) phonon DOS has the expression as

$$g(\omega) = \frac{V}{2\pi^2} \left(\frac{1}{c_l^3} + \frac{2}{c_t^3} \right) \omega^2; \tag{8.31}$$

(ii) to guarantee the total number of states satisfies

$$\int_0^{\omega_D} g(\omega)d\omega = 3nN, \tag{8.32}$$

the cut-off angular frequency called the Debye frequency ω_D is

$$\omega_D = \left(\frac{6\pi^2 nN}{V} \right)^{1/3} v_{mean} \tag{8.33}$$

where

$$\frac{3}{v_{mean}^3} = \frac{1}{c_l^3} + \frac{2}{c_t^3}; \tag{8.34}$$

(iii) the volume specific heat of the lattice is

$$C_V \triangleq \frac{\partial}{\partial T} \left[\int_0^{\omega_D} g(\omega_j(\boldsymbol{k}))\hbar\omega_j(\boldsymbol{k}) f_{\boldsymbol{k},j}^{eq} d\omega \right]$$

$$= 9nNk_B \left(\frac{T}{T_D} \right)^3 \int_0^{\frac{T_D}{T}} \frac{x^4 e^x}{(e^x - 1)^2} dx \tag{8.35}$$

where T_D is the Debye temperature defined as

$$T_D = \frac{\hbar\omega_D}{k_B}. \tag{8.36}$$

Solutions

1. **Solution**: We have obtained the expression of thermal conductivity in wave
   vector $\boldsymbol{k}$ space

$$\lambda(T) = \frac{1}{3} \frac{1}{(2\pi)^3} \sum_j \int (v_j(\boldsymbol{k}))^2 \tau_j(\boldsymbol{k}) C_{ph}(\omega_j(\boldsymbol{k}))d\boldsymbol{k}$$

$$= \frac{1}{3} \frac{1}{(2\pi)^3} \sum_j \int v_j(\boldsymbol{k})\ell_j(\boldsymbol{k}) C_{ph}(\omega_j(\boldsymbol{k}))d\boldsymbol{k}. \tag{8.37}$$

Note that

$$\sum_j \int d\mathbf{k} = \sum_j \int 4\pi k^2 dk$$

$$= \sum_j \int 4\pi \omega^2 (k/\omega)^2 \frac{dk}{d\omega} d\omega \qquad (8.38)$$

$$= \sum_j \int 4\pi \omega^2 (v_{\text{phase}})^{-2} (v_{\text{group}})^{-1} d\omega$$

and

$$g_j(\omega)/V = \frac{d \int k(\omega, j)^2 dk}{2\pi^2 d\omega} = \frac{d \int k^2 dk}{dk} \frac{dk}{2\pi^2 d\omega} = \frac{\omega^2}{2\pi^2 v_{\text{phase}}(\omega, j)^2} \frac{1}{v_{\text{group}}(\omega, j)}. \qquad (8.39)$$

Since $v_{\text{group}}(\omega, j)$ is just $v_j^2(\omega_j)$ in previous text, finally we have

$$\lambda(T) = \frac{1}{12\pi^4 V} \sum_j \int g(\omega_j) v_j^2(\omega_j) \tau_j(\omega_j) C_{\text{ph}}(\omega_j) d\omega. \qquad (8.40)$$

2. **Solution**:
   (i) for each mode, DOS is

$$g_j(\omega) d\omega = \frac{V}{(2\pi)^3} 4\pi k^2 dk = \frac{V}{(2\pi)^3} 4\pi \left(\frac{\omega}{v}\right)^2 \frac{d\omega}{v}. \qquad (8.41)$$

In Debye model, we have one longitudinal wave and two transverse waves, so we can write

$$g(\omega) = \frac{V}{2\pi^2} \left(\frac{1}{c_l^3} + \frac{2}{c_t^3}\right) \omega^2, \qquad (8.42)$$

and thus the effective mean group speed v_{mean} can de defined by

$$\frac{3}{v_{\text{mean}}^3} = \frac{1}{c_l^3} + \frac{2}{c_t^3}; \qquad (8.43)$$

(ii) Since

$$\int_0^{\omega_D} g(\omega) d\omega = 3nN, \qquad (8.44)$$

we have

$$\int_0^{\omega_D} \frac{V}{2\pi^2} \frac{3}{v_{\text{mean}}^3} \omega^2 d\omega = 3nN, \qquad (8.45)$$

so we can see

$$\omega_D = \left(\frac{6\pi^2 n N}{V}\right)^{1/3} v_{mean}. \tag{8.46}$$

(iii) The total thermal energy is

$$U = \int d\omega g(\omega) f_{k,j}^{eq} \hbar\omega = \sum_j \int_0^{\omega_D} d\omega \frac{V\omega^2}{2\pi^2 v_{group}^3} \frac{\hbar\omega}{e^{\hbar\omega/k_B T} - 1} \tag{8.47}$$

or using the mean group velocity is

$$\begin{aligned} U &= \frac{3V\hbar}{2\pi^2 v_{mean}^3} \int_0^{\omega_D} d\omega \frac{\omega^3}{e^{\hbar\omega/k_B T} - 1} \\ &= \frac{3V k_B^4 T^4}{2\pi^2 v_{mean}^3 \hbar^3} \int_0^{x_D} dx \frac{x^3}{e^x - 1} \\ &= 9N k_B T \left(\frac{T}{T_D}\right)^3 \int_0^{x_D} dx \frac{x^3}{e^x - 1} \end{aligned} \tag{8.48}$$

where T_D is the Debye temperature defined as

$$T_D = \frac{\hbar\omega_D}{k_B} \tag{8.49}$$

and $x = \hbar\omega/k_B T$. Then we can easily have

$$C_V = \frac{\partial U}{\partial T} = 9N k_B \left(\frac{T}{T_D}\right)^3 \int_0^{x_D} dx \frac{x^4 e^x}{(e^x - 1)^2}. \tag{8.50}$$

References

1. Maldovan, M.: Narrowlow-frequency spectrum and heat management by thermocrystals. Phys. Rev. Lett. **110**, 025902 (2013)
2. Maldovan, M.: Transition between ballistic and diffusive heat transport regimes in silicon materials. Appl. Phys. Lett. **101**, 113110 (2012)
3. Zen, N., Puurtinen, T.A., Isotalo, T.J., Chaudhuri, S., Maasilta, I.J.: Engineering thermal conductance using a two-dimensional phononic crystal. Nat. Commun. **5**, 3435 (2014)

Part II
Special Theories

Chapter 9
Temperature-Dependent Transformation Thermotics for Thermal Conduction: Switchable Cloak and Macroscopic Diode

Abstract By establishing temperature-dependent transformation thermotics for treating materials whose conductivity depends on temperature, we show analytical and simulation evidences for switchable thermal cloaking and a macroscopic thermal diode based on the cloaking. The latter allows heat flow in one direction but prohibits the flow in the opposite direction, which is also confirmed by our experiments. Our results suggest that the temperature-dependent transformation thermotics could be useful for achieving macroscopic heat rectification, and provide guidance both for macroscopic control of heat flow and for the design of the counterparts of switchable thermal cloaks or macroscopic thermal diodes in other fields like seismology, acoustics, electromagnetics, or matter waves.

Keywords Temperature-dependent coordinate transformation · Thermal cloak · Thermal diode

9.1 Opening Remarks

In 2008, Fan et al. [1] adopted a coordinate transformation approach to propose a class of thermal metamaterial where heat is caused to flow around an "invisible" region at steady state, thus called *thermal cloaking*. The cloaking originates from the fact that the thermal conduction equation remains form-invariant under coordinate transformation. So far, the theoretical proposal of steady-state thermal cloaking [1] and its extensions (say, bifunctional cloaking of heat and electricity [2] or nonsteady-state thermal cloaking [3]) have been experimentally verified and developed [4–8]. The theoretical treatment based on coordinate transformation [1–3, 9–13], which is called *transformation thermotics (or transformation thermodynamics)*, has a potential to become a fundamental theoretical method for macroscopically manipulating heat flow at will.

However, in order to realize desired macroscopic thermal rectification, say, switchable thermal cloaking and macroscopic thermal diodes, the existing transformation thermotics [1–3] is not enough since it only holds for materials whose conductivity is independent of temperature (thus called *linear materials*). For instance, the

© Springer Nature Singapore Pte Ltd. 2020   97
J.-P. Huang, *Theoretical Thermotics*,
https://doi.org/10.1007/978-981-15-2301-4_9

desired thermal diode should conduct heat in one direction, but insulate the heat in the opposite direction. This is a kind of asymmetric behavior of heat current. For this purpose, nonlinear materials (whose conductivity relies on temperature) must be adopted. Actually, it is long known that for many materials, their thermal conductivities (κ) vary with temperature (T): (1) κ *increases as T increases.* Say, for noncrystalline solids, a series of experiments on glass [14] showed that at low temperature the thermal conductivity is proportional to $T^{1.6} \sim T^{1.8}$; (2) κ *increases as T decreases.* For example, measurements on single crystals of silicon and germanium from 3 K to their melting point [15] showed that their thermal conductivity decreases faster than $1/T$.

9.2   Temperature-Dependent Transformation Thermotics for Thermal Conduction

Now we are in a position to develop a theory of transformation thermotics for treating nonlinear materials, thus called *temperature-dependent transformation thermotics.* The details of the theory are given in Sect. 2.3, see Eq. (2.47). Further, we obtain the following key formula,

$$\tilde{\kappa}(T) = \frac{\tilde{J}(T)\kappa_0 \tilde{J}^{t}(T)}{\det[\tilde{J}(T)]}. \tag{9.1}$$

This Eq. (9.1) is equivalent to Eq. (2.47), and it denotes that instead of constructing materials for a background whose thermal conductivity is T-dependent, we may apply a T-involved transformation to the original thermal conduction equation. This process allows us to design switchable thermal cloaks and then macroscopic thermal diodes. The former serve as an extension of the extensively investigated cloaks without switches [1–8, 16]; the latter are actually a useful application of the former in this chapter. We proceed as follows.

9.3   Switchable Thermal Cloak

9.3.1   Design

A traditional thermal cloak can protect a central region from an external heat flux and permits the region to remain a constant temperature without disturbing the temperature distribution outside the cloak (Fig. 9.1). To achieve this cloaking effect, a simple radial stretch transformation of polar coordinates may be performed. As schematically shown in Fig. 9.1, the circular region with radius R_2 is compressed to the annulus region with radius between R_1 and R_2, and the geometrical transformation can be written as

Fig. 9.1 Schematic graph depicting a thermal cloak between radius R_1 and R_2. The two lines, each with three arrows, denote the flow of heat: the cloak does not disturb the heat flow at the region with radius larger than R_2; the heat flux cannot enter the central region with radius smaller than R_1. Adapted from Ref. [25]

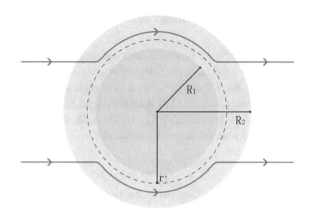

$$r' = r\frac{R_2 - R_1}{R_2} + R_1, \tag{9.2}$$

where $r \in [0, R_2]$ and $r' \in [R_1, R_2]$. Here r' is the radial coordinate in physical space.

In order to realize different responses to heat flow on the two sides of a thermal diode, we need two types of thermal cloaks: one functions at high temperature (hereafter indicated as type-A cloaks), and the other works at low temperature (type-B cloaks). Unlike some previous proposed switchable electromagnetic cloaks [17–19], the switching effect should be triggered automatically by temperature changes. For this purpose, an idea is to modify Eq. (9.2) as

$$r' = r\frac{R_2 - \tilde{R}_1(T)}{R_2} + \tilde{R}_1(T), \tag{9.3}$$

where $\tilde{R}_1(T) = R_1[1 - (1 + e^{\beta(T-T_c)})^{-1}]$ for type-A cloaks and $\tilde{R}_1(T) = R_1/(1 + e^{\beta(T-T_c)})$ for type-B cloaks. Here T_c is a critical temperature around which the cloak is switched on or off, and β is a scaling coefficient which is set to be $2.5\,\mathrm{K}^{-1}$ in this chapter.

So far, for obtaining thermal cloaks with switching phenomena, we need to combine Eqs. (9.3) and (9.1). As a result, for the area with radius $r' \in [R_1, R_2]$ in Fig. 9.1, we achieve the thermal conductivities in polar coordinates, diag$[\tilde{\kappa}_r(T), \tilde{\kappa}_\theta(T)]$, as

$$\tilde{\kappa}_r(T) = \kappa_0\frac{r' - \tilde{R}_1(T)}{r'}, \quad \tilde{\kappa}_\theta(T) = \kappa_0\frac{r'}{r' - \tilde{R}_1(T)}. \tag{9.4}$$

Here κ_0 represents the T-independent thermal conductivity of the background.

9.3.2 Finite-Element Simulation

Then we perform finite-element simulations based on the commercial software COMSOL Multiphysics (http://www.comsol.com/). A thermal cloak with $R_1 = 1$ cm and $R_2 = 2$ cm is set in a box with size 8 cm × 7 cm as shown in Fig. 9.2. Heat diffuses from the left boundary with high temperature T_H to the right boundary with low temperature T_L. Meanwhile, the upper and lower boundaries of the simulation box are thermally isolated.

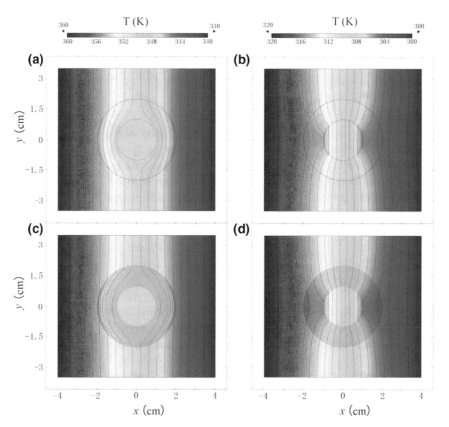

Fig. 9.2 Switchable thermal cloaks obtained by two-dimensional finite-element simulations: **a**, **c** switch on for the temperature above 340 K and **b**, **d** switch off for the temperature below 320 K. The color surface denotes the distribution of temperature, where isothermal lines are indicated; heat diffuses from left to right; the upper and lower boundaries are thermal insulation. **a** and **b** show the results of thermal conductivities calculated according to Eq. (9.4); **c** and **d** show the results of 10 alternating layers of two sub-layers with $\kappa_1(T)$ and $\kappa_2(T)$ given by Eq. (9.5) (effective medium theory). In **a**–**d**, an object with thermal conductivity 0.01 W/mK is set in the central region with radius R_1. Parameters: $\kappa_0 = 1$ W/mK, $\kappa_a = 0.1$ W/mK, $\kappa_b = 10$ W/mK, $R_1 = 1$ cm, $R_2 = 2$ cm, and $T_c = 330$ K. Adapted from Ref. [25]

The simulation results of a type-A cloak are shown in Fig. 9.2a, b. In Fig. 9.2a, at high temperature ($T = 340\,\text{K} \sim 360\,\text{K}$), we observe that the cloak is functioning and thermally hiding the object located at the central region with radius R_1. However, when the environment is changed to low temperature ($T = 300\,\text{K} \sim 320\,\text{K}$), the cloak is turned off. That is, the temperature distribution outside the object is distorted, just as the cloak (located between R_1 and R_2) is absent. Owing to the antisymmetry between type-A and type-B cloaks, the type-B cloaks exhibit the behavior similar to Fig. 9.2a, b, but switching on (or off) at low (or high) temperature.

9.3.3 Theoretical Realization Based on an Effective Medium Theory

The materials designed according to Eq. (9.4) is anisotropic and inhomogeneous, which is difficult to be realized in experiments. In fact, for constructing such materials, we can simply utilize alternating layers of two homogeneous isotropic sub-layers with thicknesses d_1 and d_2 and conductivities κ_a and κ_b. For simplification, in this chapter, we set $d_1 = d_2$. According to the theoretical analysis and effective medium theory [2–4, 16, 20], the conductivities of two sub-layers should satisfy $\kappa_a \kappa_b \approx \kappa_0^2$ for a traditional cloak. In order to endue conventional cloaks with the switching effect, some mathematical operations must be carried out on κ_a and κ_b in the way analogous to what we did on $\tilde{R}_1(T)$. Therefore, we obtain $\kappa_1(T)$ and $\kappa_2(T)$ as the new temperature-dependent thermal conductivities of the sub-layers,

$$\kappa_1(T) = \kappa_a + \frac{\kappa_0 - \kappa_a}{1 + e^{\beta(T - T_C)}}, \kappa_2(T) = \kappa_b - \frac{\kappa_b - \kappa_0}{1 + e^{\beta(T - T_C)}}. \tag{9.5}$$

The two expressions yield: $\kappa_1(T) \to \kappa_a$ and $\kappa_2(T) \to \kappa_b$ when $T \gg T_C$; $\kappa_1(T) \to \kappa_0$ and $\kappa_2(T) \to \kappa_0$ as $T \ll T_C$. That is, the cloaking effect is switched on (or off) for high (or low) temperature environment $T \gg T_C$ (or $T \ll T_C$). Equation (9.5) offers a convenient tool to help experimentally realize our theoretical design of Eq. (9.4). Next, we plot Fig. 9.2c, d, which shows the simulation results of 10 alternating layers for constructing type-A cloaks. Evidently, Fig. 9.2c, d displays the switching phenomenon similar to Fig. 9.2a, b. The procedure holds the same for achieving type-B cloaks.

9.4 Macroscopic Thermal Diode

9.4.1 Design

The above thermal cloaking may help to design a kind of macroscopic thermal diodes. As shown in Fig. 9.3a, the diode device contains Regions I, II and III: Region I (II) is a segment of the type-A (type-B) cloak, and Region III is a thermal conductor.

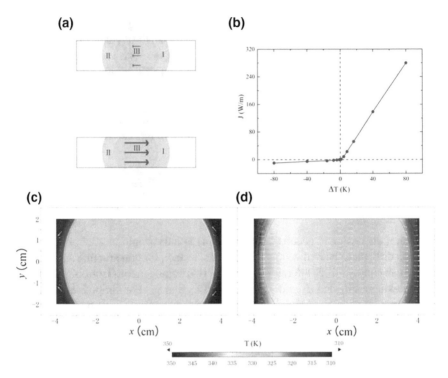

Fig. 9.3   a Sketch of a thermal diode, which is the rectangular area enclosed by the solid black lines. The blurred area outside is a reference and actually does not exist in the design. I, II, and III represent three regions, respectively. Here the arrows indicate the direction of heat flow; the length of arrows represents the amount of heat flux: the heat flux transferred from right to left (upper panel: the insulating case) is much smaller than that from left to right (lower panel: the conducting case). **b** Heat current J versus temperature bias ΔT. **c, d** Thermal diode obtained by two-dimensional finite-element simulations: **c** the insulating case and **d** the conducting case. The color surface denotes the distribution of temperature; white arrows represent the direction of heat flow; the length of white arrows indicates the amount of heat flux; the upper and lower boundaries are thermal insulation. Thermal conductivities are calculated according to Eq. (9.4); an object with thermal conductivity 10 W/mK is set in the central region with radius R_1. Parameters: $\kappa_0 = 1$ W/mK, $R_1 = 3.6$ cm, $R_2 = 4$ cm, and $T_c = 330$ K. Adapted from Ref. [25]

Compared with a full system, the cloaking effect still exists in our diode but is not perfect. There will be a small amount of heat flux conducting through the central region for the insulating case. However, the truncation of a whole cloak is necessary to separate the type-A part and the type-B part. The antisymmetry of type-A and type-B cloaks is expected to cause different behaviors of heat conducting from the two opposite directions. The transformation plays an important role in introducing anisotropic effect to the structure, which guides the heat flux to the boundaries to enhance the thermal insulating effect.

9.4.2 Finite-Element Simulation

Then we perform finite-element simulations. Figure 9.3c, d shows the simulation results of the device, which helps to insulate heat from right to left but conduct the heat from left to right. That is, the behavior of diode can be achieved indeed due to the antisymmetry of type-A and type-B cloaks (namely, Region I and Region II). Moreover, for different temperature biases (obtained by subtracting the temperature at the left boundary from that at the right boundary), ΔT, we also calculate the total heat current J by integrating the x component of heat flux across the line $x = 0$; see Fig. 9.3b. The device displays a significant rectifying effect, which has a maximum rectification ratio of 30 for the current parameter set.

9.4.3 Experimental Realization Based on an Effective Medium Theory

In order to realize such a macroscopic thermal diode, we can also adopt the effective medium theory. As discussed above, the switchable thermal cloak can be constructed with alternating layers. The thermal conductivities of the sub-layers are required to be sensitive to the temperature change around the critical point. This kind of behaviors can be found in the phase transitions of some materials [21–23]. However, inspired by the spirit of metamaterials (yielding novel functions by assembling conventional materials into specific structures), we manage to realize the macroscopic thermal diode with materials of constant conductivities. Our method is that instead of directly resorting to the transitions of materials' physical properties, we use the structural transition to trigger the switching effect. According to the demands of our design, the geometrical configuration of the device should change rapidly as temperature varies. Fortunately, we found that the shape-memory alloy (SMA) [24] may be able to help. As shown in the schematic diagrams of our design [see Fig. 9.4a and b; more details of the experiment can be found in Part II of Supplemental Material of Ref. [25]], the sub-layers of the cloak segments are coppers and expanded-polystyrene (EPS). Around the critical temperature, the deformations of the SMA slices drive the copper slices to connect or disconnect the copper layers. The connection and disconnection can be equivalently regarded as transitions of the local thermal conductivities. Thus a temperature-dependent thermal metamaterial is realized with materials of constant thermal conductivities (a whole switchable thermal cloak can also be built with the same method). The experimental results are shown in Fig. 9.4c and d. Figure 9.4c displays the temperature distribution within the central region, which is almost constant. That is, in this case, heat almost cannot flow through this central region. Thus, Fig. 9.4c corresponds to the insulating case of the diode. In contrast, Fig. 9.4d represents the conducting case of the diode. This is because Fig. 9.4d shows a significant temperature gradient within the central region. Also, the temperature distribution appears to be horizontal. As a result, we can conclude that an evident heat flow comes to appear within the central region of Fig. 9.4d.

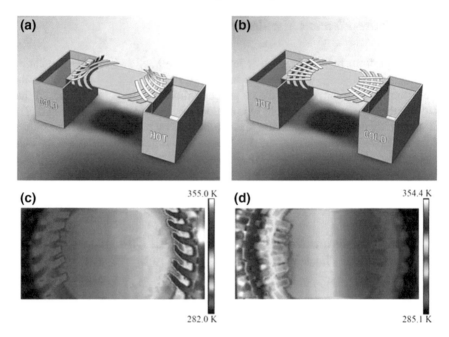

Fig. 9.4 **a**, **b** Scheme of experimental demonstration of the macroscopic thermal diode: **a** insulating case and **b** conducting case. Both the copper-made concentric layered structure and the central copper plate (both displayed in orange) are placed on an expanded-polystyrene (EPS) plate which is not shown for clarity. The left and right sides of the diode are stuck in water to have constant temperature boundary conditions. **a** When cold water is filled in the left container (light blue) and hot water the right container (pale red), the bimetallic strips of SMA and copper (white) warp up and the device blocks heat from right to left. **b** When the two containers swap their locations, the bimetallic strips (white) flatten and the device conducts heat from left to right. **c**, **d** Experimentally measured temperature distribution of the device: **c** insulating case and **d** conducting case. Adapted from Ref. [25]

9.5   Conclusions

We have established a theory of temperature-dependent transformation thermotics for dealing with thermal materials whose conductivity is temperature-dependent. The theory serves as a fundamental theoretical method for designing switchable thermal cloaking. We have also shown that the switchable thermal cloaks can be employed for achieving a macroscopic thermal diode, which has also been experimentally realized by assembling homogeneous and isotropic materials according to the design based on the effective medium theory. The diode has plenty of potential applications related to heat preservation, heat dissipation, or even heat illusion [26, 27] in many areas like efficient refrigerators, solar cells, energy-saving buildings, and military camouflage. Thus, by using temperature-dependent transformation thermotics to tailor nonlinear effects appropriately, it becomes possible to achieve desired thermal metamaterials with diverse capacities for macroscopic thermal rectification. On the same footing, our consideration (for cloaks and diodes) adopted in this chapter can be extended to obtain the counterparts of both switchable thermal cloaks and macroscopic thermal diodes in other fields like seismology, acoustics, electromagnetics, or matter waves.

9.6 Exercises and Solutions

Exercises

1. Prove both Eq. (9.1) and its equivalence with Eq. (2.47).

Solutions

1. **Solution:** See Part I of Supplemental Material of the article "Y. Li, X. Y. Shen, Z. H. Wu, J. Y. Huang, Y. X. Chen, Y. S. Ni, and J. P. Huang, Temperature-dependent transformation thermotics: From switchable thermal cloaks to macroscopic thermal diodes, Phys. Rev. Lett. **115**, 195503 (2015)", where Eq. (S14) and its proof serve as the solution.

References

1. Fan, C.Z., Gao, Y., Huang, J.P.: Shaped graded materials with an apparent negative thermal conductivity. Appl. Phys. Lett. **92**, 251907 (2008)
2. Li, J.Y., Gao, Y., Huang, J.P.: A bifunctional cloak using transformation media. J. Appl. Phys. **108**, 074504 (2010)
3. Guenneau, S., Amra, C., Veynante, D.: Transformation thermodynamics: cloaking and concentrating heat flux. Opt. Express **20**, 8207–8218 (2012)
4. Narayana, S., Sato, Y.: Heat flux manipulation with engineered thermal materials. Phys. Rev. Lett. **108**, 214303 (2012)
5. Schittny, R., Kadic, M., Guenneau, S., Wegener, M.: Experiments on transformation thermodynamics: molding the flow of heat. Phys. Rev. Lett. **110**, 195901 (2013)
6. Xu, H.Y., Shi, X.H., Gao, F., Sun, H.D., Zhang, B.L.: Ultrathin three-dimensional thermal cloak. Phys. Rev. Lett. **112**, 054301 (2014)
7. Han, T.C., Bai, X., Gao, D.L., Thong, J.T.L., Li, B.W., Qiu, C.-W.: Experimental demonstration of a bilayer thermal cloak. Phys. Rev. Lett. **112**, 054302 (2014)
8. Ma, Y.G., Liu, Y.C., Raza, M., Wang, Y.D., He, S.L.: Experimental demonstration of a multiphysics cloak: manipulating heat flux and electric current simultaneously. Phys. Rev. Lett. **113**, 205501 (2014)
9. Leonhardt, U.: Optical conformal mapping. Science **312**, 1777–1780 (2006)
10. Pendry, J.B., Schurig, D., Smith, D.R.: Controlling electromagnetic fields. Science **312**, 1780–1782 (2006)
11. Chen, H., Chan, C.T., Sheng, P.: Transformation optics and metamaterials. Nat. Mater. **9**, 387–396 (2010)
12. Landy, N., Smith, D.R.: A full-parameter unidirectional metamaterial cloak for microwaves. Nat. Mater. **12**, 25–28 (2013)
13. Kadic, M., Bückmann, T., Schittny, R., Wegener, M.: Metamaterials beyond electromagnetism. Rep. Prog. Phys. **76**, 126501 (2013)

14. Zeller, R.C., Pohl, R.O.: Thermal conductivity and specific heat of noncrystalline solids. Phys. Rev. B **4**, 2029–2041 (1971)
15. Glassbrenner, C.J., Slack, G.A.: Thermal conductivity of silicon and germanium from 3 K to the melting point. Phys. Rev. **134**, A1058–A1069 (1964)
16. Han, T.C., Yuan, T., Li, B.W., Qiu, C.-W.: Homogeneous thermal cloak with constant conductivity and tunable heat localization. Sci. Rep. **3**, 1593 (2013)
17. Chen, P.Y., Alù, A.: Atomically thin surface cloak using graphene monolayers. ACS Nano **5**, 5855–5863 (2011)
18. Zhang, W., Zhu, W.M., Cai, H., Tsai, M.-L.J., Lo, G.Q., Tsai, D.P., Tanoto, H., Teng, J.H., Zhang, X.H., Kwong, D.L., Liu, A.Q.: Resonance switchable metamaterials using MEMS fabrications. IEEE J. Sel. Top. Quant. **19**, 4700306 (2013)
19. Wang, R.F., Mei, Z.L., Yang, X.Y., Ma, X., Cui, T.J.: Switchable invisibility cloak, anticloak, transparent cloak, superscatterer, and illusion for the laplace equation. Phys. Rev. B **89**, 165108 (2014)
20. Huang, J.P., Yu, K.W.: Enhanced nonlinear optical responses of materials: composite effects. Phys. Rep. **431**, 87–172 (2006)
21. Oh, D.-W., Ko, C., Ramanathan, S., Cahill, D.G.: Thermal conductivity and dynamic heat capacity across the metal-insulator transition in thin film VO_2. Appl. Phys. Lett. **96**, 151906 (2010)
22. Zheng, R.T., Gao, J.W., Wang, J.J., Chen, G.: Reversible temperature regulation of electrical and thermal conductivity using liquid-solid phase transitions. Nat. Commun. **2**, 289 (2011)
23. Siegert, K.S., Lange, F.R.L., Sittner, E.R., Volker, H., Schlockermann, C., Siegrist, T., Wuttig, M.: Impact of vacancy ordering on thermal transport in crystalline phase-change materials. Rep. Prog. Phys. **78**, 013001 (2015)
24. Omori, T., Kainuma, R.: Materials science: alloys with long memories. Nature **502**, 42–44 (2013)
25. Li, Y., Shen, X.Y., Wu, Z.H., Huang, J.Y., Chen, Y.X., Ni, Y.S., Huang, J.P.: Temperature-dependent transformation thermotics: from switchable thermal cloaks to macroscopic thermal diodes. Phys. Rev. Lett. **115**, 195503 (2015)
26. Han, T.C., Bai, X., Thong, J.T.L., Li, B.W., Qiu, C.-W.: Full control and manipulation of heat signatures: cloaking, camouflage and thermal metamaterials. Adv. Mat. **26**, 1731–1734 (2014)
27. Zhu, N.Q., Shen, X.Y., Huang, J.P.: Converting the patterns of local heat flux via thermal illusion device. AIP Adv. **5**, 053401 (2015)

Chapter 10
Temperature Trapping Theory: Energy-Free Thermostat

Abstract It is crucial to maintain constant temperatures in an energy-efficient way. Here we present a temperature-trapping theory for asymmetric phase-transition materials with thermally responsive thermal conductivities. Then we theoretically introduce and experimentally demonstrate a concept of energy-free thermostat within ambient temperature gradients. The thermostat is capable of self-maintaining a desired constant temperature without the need of consuming energy even though the environmental temperature gradient varies in a large range. As a model application of the concept, we design and show a different type of thermal cloak that has a constant temperature inside its central region in spite of the changing ambient temperature gradient, which is in sharp contrast to all the existing thermal cloaks. This chapter has relevance to energy-saving heat preservation, and it provides guidance both for manipulating heat flow without energy consumption and for designing new metamaterials with temperature-responsive or field-responsive parameters in many disciplines such as thermotics, optics, electromagnetics, acoustics, mechanics, electrics, and magnetism.

Keywords Thermostat · Thermal cloak · Temperature trapping

10.1 Opening Remarks

It is known that humans are faced with a global energy crisis, namely, an increasing shortage of non-renewable energy resources, such as coal, petroleum, and natural gas [1]. However, much of the energy generated from non-renewable energy resources is used for temperature preservation in many areas ranging from industrial fields to our daily lives. Therefore, it is meaningful and challenging to reduce such energy consumption.

Heat conduction exists in matter as long as there is a temperature gradient. In recent years, researchers have been devoted to understanding and controlling the conduction of heat, putting a particular emphasis on its nonlinear feature. Their fundamental interests focus on the nonlinear conduction phenomenon at the microscopic scale to have a better understanding [2, 3], to improve thermoelectric effects

© Springer Nature Singapore Pte Ltd. 2020
J.-P. Huang, *Theoretical Thermotics*,
https://doi.org/10.1007/978-981-15-2301-4_10

[4–6], or to achieve novel thermal rectification [7–9]. However, the nonlinear heat conduction at the macroscopic scale is seldom touched in the field of fundamental research even though it was already reported long time ago [10]. In this chapter, by tailoring the nonlinear effect of macroscopic heat conduction [10, 11], namely, manipulating thermally responsive thermal conductivities appropriately, we establish a temperature-trapping theory and then propose a novel concept of energy-free thermostat. The thermostat can self-maintain a desired constant temperature without the need of consuming energy even though the environmental temperature gradient changes in a large range. As a proof of concept, we experimentally fabricate a prototype device by assembling commercially available materials according to a multistep approximation method, which enables us to effectively realize the desired thermal conductivities on the same footing as thermal metamaterials [11–18] or metamaterials in other fields [19–22]. Then we apply the thermostat concept to thermal cloaking [11–18] as a model application, and show an improved thermal cloak whose central region serves as an ideal thermal environment with a constant temperature even though the environmental temperature gradient varies significantly. This feature makes the improved thermal cloak distinctly different from the existing thermal cloaks [11–18].

10.2   Temperature-Trapping Theory: Concept of Energy-Free Thermostat

It is known that thermal conductivities essentially depend on temperature, T [10]. Particularly, in a phase-transition process, the thermal conductivity can change sharply [23]. For simplicity, let us consider a one-dimensional steady-state heat conduction along x-direction with temperature-dependent thermal conductivities, $\kappa(x, T)$. The conduction follows the differential equation,

$$\frac{d}{dx}\left(\kappa(x, T)\frac{dT}{dx}\right) = \frac{\partial \kappa(x, T)}{\partial x}\frac{dT}{dx} + \frac{\partial \kappa(x, T)}{\partial T}\left(\frac{dT}{dx}\right)^2 + \kappa(x, T)\frac{d^2 T}{dx^2} = 0.$$

(10.1)

Then we are in a position to find a specific value or function of $\kappa(x, T)$ for our purpose: the temperature at a specific region should keep (almost) unchanged even though the associated boundary conditions change significantly. In other words, we expect that the desired material should be able to automatically maintain a constant temperature within changing ambient temperature gradients without the need of adding additional work.

For more details, let us consider the one-dimensional model, where the right-hand (left-hand) side of the model is the heat (cold) source with a fixed high (low) temperature of T_H (T_L). We set T_c as an arbitrary value between T_H and T_L. As T_H increases (or T_L decreases), we expect that the temperature at the middle point (located between the hot and cold sources) should always be T_c. Now, suppose the model is divided into two parts from the middle point: the temperature of Part I ranging from T_H to T_c, and Part II from T_c to T_L. In case of increasing T_H only, for any $\kappa(x, T)$, the temperature in Part II should always range from T_c to T_L. Therefore, in Part II, the temperature gradient remains unchanged, yielding a corresponding heat flow. Since T_H may increase and it has no effect on the temperature at the middle point, the heat flow must be independent of T_H. Owing to the continuity of heat flow, we obtain

$$q = -\kappa(x, T)\frac{dT}{dx} \equiv C, \tag{10.2}$$

where q is the density of heat flow, and C is a constant. Clearly, if only considering a decreasing T_L, we may obtain a same equation as Eq. (10.2). This fact means that our model is centrosymmetric: $\kappa(x, T)$ is symmetrical with respect to $T = T_c$. Equation (10.2) is the key for giving a rough sketch of $\kappa(x, T)$. As $T \to T_c$, $dT/dx \to 0$. Thus, $\kappa(x, T)$ should tend to infinity and reach the highest value. On the contrary, for $T \gg T_c$ or $T \ll T_c$, the temperature gradients are large, so $\kappa(x, T)$ tends to zero. Moreover, the middle point with $T = T_c$ is the inflection point of function $\kappa(x, T)$. To sum up, the temperature-dependent function, $\kappa(x, T)$, should satisfy the following rules: $\kappa(x, T - T_c) = \kappa(x, T_c - T)$; $\kappa(x, T) \to \infty$ as $T \to T_c$; and $\kappa(x, T) \to 0$ as $T \gg T_c$ or $T \ll T_c$. The second requirement is not realistic for extant materials and some approximation should be adopted instead.

In consideration of mathematical simplicity for easy analysis and modeling, we design an alternative one-dimensional heat-conduction model as shown in Fig. 10.1a, where T_3, T_2, T_1, and T_0 (or x_3, x_2, x_1, and x_0) are the corresponding temperatures (or positions) along the direction of heat diffusion, where we set $x_1 - x_0 = x_3 - x_2$. Two types of nonlinear materials (type-A and type-B) are located at the left-hand and right-hand sides respectively, and a common material with a high and constant conductivity κ_η is placed in the middle region. In this model, the conductivities of type-A and B are related to the temperature, and the relations are chosen to hold the same form as the Logistic functions $L(T)$ widely used in Logistic regressions. We believe the combination of a pair of axial symmetry Logistic functions may give a good approximation of the aforementioned requirements of thermal conductivity. Furthermore, these two types of materials are much more practicable because the sigmoid curves indicated by Logistic functions are similar to those "S" shape curves in phase-transition materials describing physical properties as a function of temperature [23–25].

More explicitly, the type-A nonlinear material is designed with thermal conductivity κ_A as a conductor (or an insulator) at high (or low) temperature. An inverse behavior happens for the type-B material with thermal conductivity κ_B. For a given phase-transition temperature T_c, both κ_A and κ_B may be assumed as

$$\kappa_A = L(T_c - T) = \delta + \frac{\varepsilon e^{T-T_c}}{1 + e^{T-T_c}}, \quad \kappa_B = L(T - T_c) = \delta + \frac{\varepsilon}{1 + e^{T-T_c}}, \tag{10.3}$$

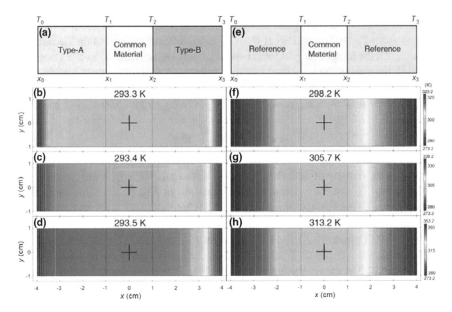

Fig. 10.1 **a** Schematic graph showing the design of energy-free thermostat, which consists of three parts: common material (2 cm width and 2 cm height), type-A and type-B nonlinear materials (each with 3 cm width and 2 cm height). **b–d** Finite-element simulations of (**a**) according to Eq. (10.3) where $\delta = 0.4$ W/m K, $\varepsilon = 49.6$ W/m K, and $T_c = 293.2$ K. **e** Schematic diagram of a reference model. **f–h** Finite-element simulations of (**e**), where the thermal conductivity of the reference material (denoted by blue color in **e**) is 50 W/m K. For all the simulations (**b–d, f–h**), the temperature at the center (marked with +) is extracted and showed above each panel. The thermal conductivity of the central common material is 400 W/m K. The left boundary is kept at a constant low temperature, 273.2 K, while the right boundary is kept at different high temperatures: 323.2 K (**b, f**), 338.2 K (**c, g**), and 353.2 K (**d, h**). The upper and lower boundaries are thermally insulated. Adapted from Ref. [29]

where δ is a small value and ε is high enough. According to the Fourier law $q = -\kappa \frac{dT}{dx}$ or $q\,dx = -\kappa\,dT$ together with the geometrical relation $(x_1 - x_0 = x_3 - x_2)$ and continuity of conduction $\int_{x_0}^{x_1} q\,dx = \int_{x_2}^{x_3} q\,dx$ or $\int_{T_0}^{T_1} \kappa_A dT = \int_{T_2}^{T_3} \kappa_B dT$, we obtain (under the assumption that $e^{T_0-T_c}$ and $e^{T_c-T_3}$ are close to zero)

$$(\varepsilon + \delta)(T_3 - T_2) - \varepsilon(T_3 - T_c) - \delta(T_1 - T_0) = \varepsilon \ln\left(\frac{1 + e^{T_1-T_c}}{1 + e^{T_2-T_c}}\right). \qquad (10.4)$$

It should be remarked that the thermal conductivity in the middle region should be much higher ($\kappa_\eta \gg \varepsilon$) to make $T_1 \approx T_2$. As a result, the relation $T_1 \approx T_2 \approx T_c$ can be obtained according to Eq. (10.4). Therefore, the temperature preserved in our device is approximately the phase-transition temperature, T_c. The above discussion shows that adopting two Logistic functions is enough for energy-freely maintaining constant temperatures as ambient temperature gradients change.

Then we perform finite-element simulations according to Eq. (10.3). In Fig. 10.1, the central temperatures at the positions marked with $+$ show a strong contrast between the thermostat (Fig. 10.1b–d) and the reference system (Fig. 10.1f–h). The former are close to T_c (within 0.3 K) for the three cases. On the contrary, the latter deviate from T_c as far as 20 K when the boundary condition changes within the same range. It is worth noting that the temperature of the cold source in Fig. 10.1b–d, f–h remains constant (273.2 K) for the sake of comparison. Actually, if this temperature varies as well, the behavior of the central temperatures will keep unchanged due to the fixed value of T_c.

10.3 Experimental Demonstration of the Energy-Free Thermostat Concept

The major challenge in realizing the thermostat is to find the type-A and type-B materials with nonlinear (temperature-dependent) conductivities (Eq. (10.3)). Although phase transitions may generate such variations of conductivities, the transition process is lack of convenience in operations. We believe that a simplest temperature-dependent conductivity is to switch between two values, which can be achieved by connecting two thermal conductors or not. This process may be described by a step function. To make the conductivity vary as a step function, a practical way is to assemble a still part and a movable part. Figure 10.2 shows the relevant details. In our experiment, the still part is constructed by placing three phosphor copper (QSn6.5-0.1) films ($\kappa_p = 54$ W/m K) at regular intervals; each interval is filled with silicon grease ($\kappa_s = 4$ W/m K). All the phosphor copper films and their intervals (filled with silicon grease) hold the same dimensions. As a result, the effective thermal conductivity of this alternating layered structure can be derived as $\kappa_{eff} = \left[\left(\kappa_p^{-1} + \kappa_s^{-1}\right)/2\right]^{-1} = 7.4$ W/m K [26]. On the other hand, a bimetallic strip composed of phosphor copper and shape memory alloy (SMA) [27, 28] serves as the movable part.

The SMA is capable of changing its shape as the temperature varies. Specifically, a two-way SMA chosen for building the type-A material is able to tilt up an angle below 278.2 K and completely level above 297.2 K. For the type-B material, the transition temperature is the same but the deformation is opposite. Therefore, as we change the temperature of the hot or cold source, the bimetallic strips will be driven up and down. In general, due to the deformation of SMA, 0–3 metal films will fill the gaps between the two connective layers of phosphor copper (Fig. 10.2a–b), and change the effective conductivity of type-A or type-B material according to the effective medium theory [26]. The calculated thermal conductivities are displayed in Fig. 10.2c. It should also be noted that the transition temperature T_c of the Type-A and Type-B materials is actually 297.2 K, since at this temperature, the three SMAs are all flat (or warped) for type-A (or type-B).

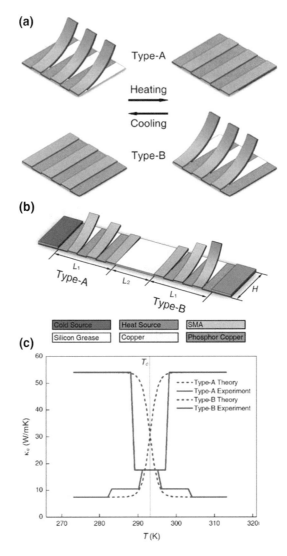

Fig. 10.2 **a** Schematic graph showing the type-A and type-B nonlinear materials of the energy-free thermostat fabricated in our experiment. For the type-A part, the bimetallic strips will tilt up and result in a low effective thermal conductivity κ_{eff} of the layered structure when the temperature is below 297.2 K. Upon heating, the bimetallic strips will become level and replace the silicon grease layers with phosphor copper when the temperature is above 297.2 K. As a result, κ_{eff} increases. The opposite behavior happens for the deformation of the type-B part. When a temperature gradient exists in the device, the bimetallic strips show different degrees of deformation; see **b**. The dimensions indicated in **b**: $L_1 = 3$ cm, $L_2 = 2$ cm, and $H = 2$ cm. Since κ_{eff} of the type-A or type-B part depends on the shapes of the bimetallic strips, which vary with the temperature T_S of the heat/cold sources, we can achieve a relationship $\kappa_{eff}(T_S)$ that is similar to $\kappa_A(T)$ or $\kappa_B(T)$ in Eq. (10.3). This similarity serves as the main principle of our experiment; see **c**. The solid curves indicated by "Experiment" are calculated with the effective medium theory by assuming the leveling of 0–3 bimetallic strips; the dashed curves indicated by "Theory" are the result of Eq. (10.3). Adapted from Ref. [29]

In addition, a copper film with $\kappa_\eta = 394\,\mathrm{W/m\,K}$ is located in the middle to assemble the whole device; see Fig. 10.2b. Moreover, all the metal surfaces are covered by polydimethylsiloxane, which helps against heat dissipation and makes the whole device "visible" for the thermal camera.

As shown in Fig. 10.3, the experiment is carried out for three different boundary conditions, where the temperature of the heat source is changed significantly while the temperature of the cold source keeps almost unchanged. Figure 10.3a–c shows the measured temperatures at the three centers, which are almost constant as predicted by the theory. The constant value is a little lower than the phase-transition temperature (297.2 K) due to the dissipation of heat to the environment. Meanwhile, the temperature distributions of the device without the SMA are also presented in Fig. 10.3d–f for comparison. In this case, the temperatures within the central areas vary significantly (Fig. 10.3d–f).

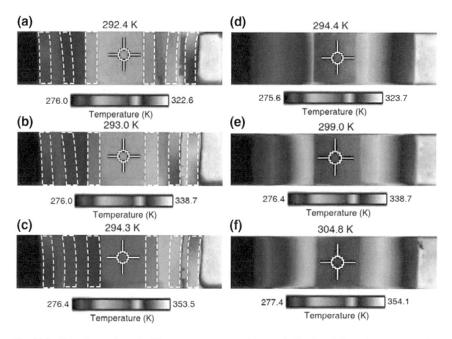

Fig. 10.3 Experimental results. The temperature at each center is displayed above the corresponding panel. For the energy-free thermostat (**a–c**), the bimetallic strips are indicated with white dashed lines. As the temperature of the heat source varies in a wide range (322.6–353.5 K), the central temperature keeps almost unchanged (**a–c**). In contrast, for the reference system (**d–f**), the measured central temperature varies evidently. Adapted from Ref. [29]

10.4 Apply the Energy-Free Thermostat Concept to Design a New Thermal Cloak

To make our thermostat concept more useful, we try applying it to thermal cloaking [11–18]. As a result, we shall achieve an improved thermal cloak whose central region serves as an ideal thermal environment with a constant temperature even though the environmental temperature gradient changes significantly. This feature makes the improved thermal cloak distinctly different from the existing thermal cloaks (whose central temperatures vary evidently as environmental temperature gradients change significantly) [11–18].

A thermal cloak helps to steer heat flow around an object without the need of disturbing the temperature distribution outside the object, and it has two basic characteristics: an undisturbed temperature field outside and a uniform temperature field inside [11–18]. To design our improved thermal cloak, we resort to the structure of bilayer thermal cloak [18] by using four types of materials; see Fig. 10.4a which shows Regions I–VI for the new device in a background with thermal conductivity κ_0. Regions I–IV have a height of $2R_1$, and they are occupied by the four types of materials, whose thermal conductivities are respectively given by

$$\kappa_I = \kappa_0 - \frac{(\kappa_0 - \kappa_i)}{1 + e^{T-T_C}}, \quad \kappa_{II} = \kappa_0 - \frac{(\kappa_0 - \kappa_i)e^{T-T_C}}{1 + e^{T-T_C}},$$

$$\kappa_{III} = \kappa_0 + \frac{(\kappa_e - \kappa_0)}{1 + e^{T-T_C}}, \quad \kappa_{IV} = \kappa_0 + \frac{(\kappa_e - \kappa_0)e^{T-T_C}}{1 + e^{T-T_C}}. \tag{10.5}$$

These conductivities may be manufactured by utilizing the multistep approximation method as employed in the experimental design (Fig. 10.2c). In Eq. (10.5), κ_i and κ_e are the parameters of bilayer cloak: thermal conductivity of inner ring $\kappa_i \to 0$ and thermal conductivity of outer ring $\kappa_e = \kappa_0(R_3^2 + R_2^2)/(R_3^2 - R_2^2)$. On the other hand, Regions V and VI in Fig. 10.4a are occupied by two common materials with conductivities

$$\kappa_V \to 0 \text{ and } \kappa_{VI} = \kappa_0(R_3^2 + R_2^2)/(R_3^2 - R_2^2) \tag{10.6}$$

according to the requirement of bilayer thermal cloaks [18]. The simulation results in Fig. 10.4b–d show that as the temperature of heat source is increased from 323.2 to 338.2 K and to 353.2 K, the improved thermal cloak is able to energy-freely maintain an approximately constant temperature inside it indeed; see 293.5, 294.0 K, and 294.7 K as indicated in Fig. 10.4b–d. We obtain this conclusion by comparing the "293.5, 294.0 and 294.7 K" with their counterpart values of the common cloaking (namely, "298.2, 305.7 and 313.2 K", which are calculated by averaging the two temperatures of the cold and hot sources when the cloak is located in the center between the two sources).

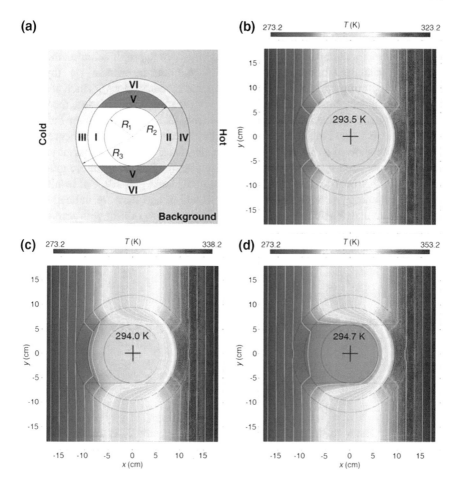

Fig. 10.4 a Schematic graph showing the improved bilayer thermal cloak in a background. The thermal conductivities in Regions I–IV (or V–VI) are determined by Eq. (10.5) (or Eq. (10.6)). **b–d** Finite-element simulations of the improved thermal cloak, where the white lines denote the isothermal lines. **b–d** Show a central circular region where the temperature keeps approximately unchanged even though the temperature of the heat source increases significantly. Parameters: $\kappa_0 = 2.3\,\text{W/m K}$, $\kappa_V = 0.03\,\text{W/m K}$, $R_1 = 6\,\text{cm}$, $R_2 = 9.5\,\text{cm}$, and $R_3 = 12\,\text{cm}$. Adapted from Ref. [29]

10.5 Discussion and Conclusions

We have established a temperature-trapping theory for asymmetric phase-transition materials with thermally responsive thermal conductivities, and introduced a concept of energy-free thermostat. The thermostat is capable of self-maintaining a desired constant temperature without energy consumption even though the environmental temperature gradient varies in a large range. (On the same footing, a so-called negative-energy thermostat [30] can also be designed by introducing thermoelectric effects, which generates electricity associated with energy-free maintenance of a constant ambient temperature.) By using a multistep approximation method, we have experimentally fabricated a prototype device. In the experiment, we employed homogenous isotropic materials and SMAs (shape memory alloys), which are commercially available. Since no heat engine or temperature sensor is used, the device is light-weighted and can be used as components of larger facilities or buildings without interference. For instance, inspired by this concept, we have theoretically designed and showed a different type of thermal cloak that has a constant temperature inside its central region in spite of changing ambient temperature gradients, which is in sharp contrast to all the existing thermal cloaks [11–18]. Our results have relevance to energy-saving temperature preservation, and they are indicative of a great freedom in extensibility, e.g., for controlling the flow of heat with zero-energy consumption and for designing new metamaterials with temperature-responsive or field-responsive parameters in many disciplines such as thermotics, optics, electromagnetics, acoustics, mechanics, electrics, and magnetism.

10.6   Exercises and Solutions

Exercises

1. Prove Eq. (10.4).

Solutions

1. **Solution**: Refer to the article "J. Wang, J. Shang, and J. P. Huang, Negative energy consumption of thermostats at ambient temperature: Electricity generation with zero energy maintenance, Phys. Rev. Appl. **11**, 024053 (2019)", where the Eqs. (1)–(10) just show the proof for Eq. (10.4).

References

1. Chu, S., Majumdar, A.: Opportunities and challenges for a sustainable energy future. Nature (London) **488**, 294–303 (2012)
2. Maldovan, M.: Phonon wave interference and thermal bandgap materials. Nature Mater. **14**, 667–674 (2015)
3. Hu, Y., Zeng, L., Minnich, A.J., Dresselhaus, M.S., Chen, G.: Spectral mapping of thermal conductivity through nanoscale ballistic transport. Nature Nanotech. **10**, 701–706 (2015)
4. Bell, L.E.: Cooling, heating, generating power, and recovering waste heat with thermoelectric systems. Science **321**, 1457–1461 (2008)
5. Biswas, K., He, J., Blum, I.D., Wu, C.-I., Hogan, T.P., Seidman, D.N., Dravid, V.P., Kanatzidis, M.G.: High-performance bulk thermoelectrics with all-scale hierarchical architectures. Nature (London) **489**, 414–418 (2012)

6. Zhao, L.D., Lo, S.-H., Zhang, Y., Sun, H., Tan, G., Uher, C., Wolverton, C., Dravid, V.P., Kanatzidis, M.G.: Ultralow thermal conductivity and high thermoelectric figure of merit in SnSe crystals. Nature (London) **508**, 373–377 (2014)

7. Li, B.W., Wang, L., Casati, G.: Thermal diode: rectification of heat flux. Phys. Rev. Lett. **93**, 184301 (2004)

8. Chang, C.W., Okawa, D., Majumdar, A., Zettl, A.: Solid-state thermal rectifier. Science **314**, 1121–1124 (2006)

9. Martínez-Pérez, M.J., Fornieri, A., Giazotto, F.: Rectification of electronic heat current by a hybrid thermal diode. Nature Nanotech. **10**, 303–307 (2015)

10. Glassbrenner, C.J., Slack, G.A.: Thermal conductivity of silicon and germanium from 3 K to the melting point. Phys. Rev. **134**, A1058–A1069 (1964)

11. Li, Y., Shen, X.Y., Wu, Z.H., Huang, J.Y., Chen, Y.X., Ni, Y.S., Huang, J.P.: Temperature-dependent transformation thermotics: from switchable thermal cloaks to macroscopic thermal diodes. Phys. Rev. Lett. **115**, 195503 (2015)

12. Fan, C.Z., Gao, Y., Huang, J.P.: Shaped graded materials with an apparent negative thermal conductivity. Appl. Phys. Lett. **92**, 251907 (2008)

13. Narayana, S., Sato, Y.: Heat flux manipulation with engineered thermal materials. Phys. Rev. Lett. **108**, 214303 (2012)

14. Maldovan, M.: Sound and heat revolutions in phononics. Nature **503**, 209–217 (2013)

15. Schittny, R., Kadic, M., Guenneau, S., Wegener, M.: Experiments on transformation thermo-dynamics: molding the flow of heat. Phys. Rev. Lett. **110**, 195901 (2013)

16. Ma, Y.G., Liu, Y.C., Raza, M., Wang, Y.D., He, S.L.: Experimental demonstration of a multiphysics cloak: manipulating heat flux and electric current simultaneously. Phys. Rev. Lett. **113**, 205501 (2014)

17. Xu, H.Y., Shi, X.H., Gao, F., Sun, H.D., Zhang, B.L.: Ultrathin three-dimensional thermal cloak. Phys. Rev. Lett. **112**, 054301 (2014)

18. Han, T.C., Bai, X., Gao, D.L., Thong, J.T.L., Li, B.W., Qiu, C.-W.: Experimental demonstration of a bilayer thermal cloak. Phys. Rev. Lett. **112**, 054302 (2014)

19. Leonhardt, U.: Optical conformal mapping. Science **312**, 1777–1780 (2006)

20. Pendry, J.B., Schurig, D., Smith, D.R.: Controlling electromagnetic fields. Science **312**, 1780–1782 (2006)

21. High, A.A., Devlin, R.C., Dibos, A., Polking, M., Wild, D.S., Perczel, J., de Leon, N.P., Lukin, M.D., Park, H.: Visible-frequency hyperbolic metasurfaces. Nature (London) **522**, 192–196 (2015)

22. Kaina, N., Lemoult, F., Fink, M., Lerosey, G.: Negative refractive index and acoustic superlens from multiple scattering in single negative metamaterials. Nature (London) **525**, 77–81 (2015)

23. Zheng, R.T., Gao, J.W., Wang, J.J., Chen, G.: Reversible temperature regulation of electrical and thermal conductivity using liquid-solid phase transitions. Nat. Commun. **2**, 289 (2011)

24. Oh, D.-W., Ko, C., Ramanathan, S., Cahill, D.G.: Thermal conductivity and dynamic heat capacity across the metal-insulator transition in thin film VO_2. Appl. Phys. Lett. **96**, 151906 (2010)

25. Siegert, K.S., Lange, F.R.L., Sittner, E.R., Volker, H., Schlockermann, C., Siegrist, T., Wuttig, M.: Impact of vacancy ordering on thermal transport in crystalline phase-change materials. Rep. Prog. Phys. **78**, 013001 (2015)

26. Huang, J.P., Yu, K.W.: Enhanced nonlinear optical responses of materials: composite effects. Phys. Rep. **431**, 87–172 (2006)

27. Chluba, C., Ge, W., de Miranda, R.L., Strobel, J., Kienle, L., Quandt, E., Wuttig, M.: Ultralow-fatigue shape memory alloy films. Science **348**, 1004–1007 (2015)

28. Dye, D.: Shape memory alloys: towards practical actuators. Nature Mater. **14**, 760–761 (2015)

29. Shen, X.Y., Li, Y., Jiang, C.R., Huang, J.P.: Temperature trapping: energy-free maintenance of constant temperatures as ambient temperature gradients change. Phys. Rev. Lett. **117**, 055501 (2016)

30. Wang, J., Shang, J., Huang, J.P.: Negative energy consumption of thermostats at ambient temperature: electricity generation with zero energy maintenance. Phys. Rev. Appl. **11**, 024053 (2019)

Chapter 11
Coupling Theory for Temperature-Independent Thermal Conductivities: Thermal Correlated Self-Fixing

Abstract It is a challenge to design intelligent thermal metamaterials due to the lack of suitable theories. Here we propose a kind of intelligent thermal metamaterials by investigating a core-shell structure, where both the core and shell have an anisotropic thermal conductivity. We solve Laplace's equation for deriving the equivalent thermal conductivity of the core-shell structure. Amazingly, the solution gives two coupling relations of conductivity tensors between the core and shell, which cause the whole core-shell structure to counter-intuitively self-fix a constant isotropic conductivity even when the area or volume fraction of core changes within the full range in two or three dimensions. The theoretical findings on fraction-independent properties are in sharp contrast to those predicted by the well-known effective medium theories, and they are further confirmed with our laboratory experiments and computer simulations. This chapter offers two coupling relations for designing intelligent thermal metamaterials, and they are not only helpful for thermal stabilization or camouflage/illusion, but also offer hints on how to achieve similar metamaterials in other fields.

Keywords Correlated self-fixing · Coupling · Core-shell structure · Intelligent thermal metamaterials

11.1 Opening Remarks

Thermal metamaterials [1, 2] are actually artificial materials or devices that exhibit novel thermal properties based on their geometrical structures or patterns. The earliest example of thermal metamaterials is thermal cloaks designed by coordinate transformation, which guide the heat flow around an object as if the object does not exit [3–8]. In principle, thermal metamaterials are useful to efficiently control the heat flow. However, almost all the existing thermal metamaterials are non-intelligent, which means that the corresponding thermal metamaterials cannot feel and respond to the change of external or internal stimuli in a controlled fashion. On the contrary, if such metamaterials can do so, they can be called intelligent thermal metamaterials.

© Springer Nature Singapore Pte Ltd. 2020
J.-P. Huang, *Theoretical Thermotics*,
https://doi.org/10.1007/978-981-15-2301-4_11

Owing to the lack of suitable theories, intelligent thermal metamaterials have not been touched in the literature, except for those with dual functions [9–15] or with thermal responsiveness [16, 17]. For example, in ref. [16], Li et al. developed a theory of temperature-dependent transformation thermotics, and designed an intelligent thermal cloak. The cloak can feel the change of environmental temperatures, and then it can be automatically switched on or off. Nevertheless, this kind of intelligent thermal cloaks feel and respond to a kind of external stimuli (namely, the change of environmental temperatures), rather than internal stimuli (e.g., the change of internal states). In fact, for external stimuli, similar intelligence appears in all the other existing intelligent thermal metamaterials [9–17]. Regarding the internal stimuli, no work has been reported to date. Thus, in this chapter, we start to propose a class of intelligent thermal metamaterials that can feel and respond to a type of internal stimuli, namely, the change of area/volume fractions. In practice, area or volume fractions can be changed due to either expansion caused by heat or contraction caused by cold. For simplicity, this chapter focuses only on the cases corresponding to the pure change of area or volume fractions in a core-shell structure, where both the core and shell have an anisotropic thermal conductivity.

Then we theoretically reveal two coupling relations for counter-intuitively self-fixing a constant isotropic thermal conductivity of the core-shell structure as the area or volume fraction varies within the full range. To this end, our theoretical results (namely, fraction-independent properties) are validated by both experiments and simulations in two or three dimensions, which are in sharp contrast to the fraction-dependent properties predicted by the well-known effective medium theories including the Bruggeman formula and the Maxwell-Garnett formula. Such effective medium theories have been extensively adopted in the field of metamaterials from optics/electromagnetics, to acoustics, to mechanics, and to thermotics.

11.2 Theory for Two Dimensions

Let us start by considering a core-shell structure embedded in a host, whose center is located in the origin of coordinates. See Fig. 11.1a. The core (or shell) has radius r_1 (or r_2) and thermal conductivity κ_1 (or κ_2). κ_3 denotes the host's thermal conductivity. For the sake of generality, we assume the core and shell to be anisotropic. In cylindrical coordinates (r, θ), κ_1 and κ_2 are both tensorial, which can be represented by diag $\left(\kappa_{rr_1}, \kappa_{\theta\theta_1}\right)$ and diag $\left(\kappa_{rr_2}, \kappa_{\theta\theta_2}\right)$, respectively.

In this case, for a passive and stable heat transport process, Laplace's equation has the following form

$$\frac{1}{r}\frac{\partial}{\partial r}\left(r\kappa_{rr}\frac{\partial T}{\partial r}\right) + \frac{1}{r}\frac{\partial}{\partial\theta}\left(\frac{\kappa_{\theta\theta}}{r}\frac{\partial T}{\partial\theta}\right) = 0, \qquad (11.1)$$

where T denotes temperature. The general solution of Eq. (11.1) is given by

$$T = A_0 + B_0 \ln r$$

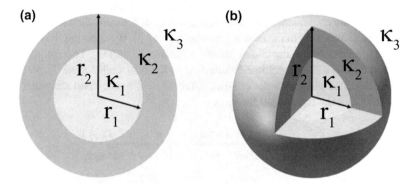

Fig. 11.1 Schematic graph showing the core-shell structure in **a** two dimensions and **b** three dimensions. Adapted from Ref. [20]

$$+ \sum_{m=1}^{\infty} [A_m \cos(m\theta) + B_m \sin(m\theta)] r^m \sqrt{\frac{\kappa_{\theta\theta}}{\kappa_{rr}}}$$

$$+ \sum_{n=1}^{\infty} [C_n \cos(n\theta) + D_n \sin(n\theta)] r^{-n} \sqrt{\frac{\kappa_{\theta\theta}}{\kappa_{rr}}}. \tag{11.2}$$

It is worth noting that Eq. (11.2) is valid under the condition that both κ_{rr} and $\kappa_{\theta\theta}$ are either positive or negative. Here it is necessary for us to remark that apparently negative thermal conductivities can only be achieved by adding external work as required by the second law of thermodynamics, which mean that the heat (e.g., in electric refrigerators) can be driven to transport from a region with a low temperature to another region with a high temperature; more relevant comments can be found in the Supplementary Information.

For clarity, we use T_1, T_2 and T_3 to denote the temperature distribution in core, shell and host. To focus on the core-shell structure, we simplify the host to be homogeneous and isotropic. As a result, we obtain T_3 as

$$T_3 = A_0 + A_1 r \cos\theta + C_1 r^{-1} \cos\theta. \tag{11.3}$$

Meanwhile, this system has the corresponding boundary conditions [18],

$$\begin{cases} T_1 \text{ is finite} \\ T_1(r_1) = T_2(r_1) \\ T_2(r_2) = T_3(r_2) \\ -\kappa_{rr_1} \frac{\partial T_1}{\partial r}\Big|_{r=r_1} = -\kappa_{rr_2} \frac{\partial T_2}{\partial r}\Big|_{r=r_1} \\ -\kappa_{rr_2} \frac{\partial T_2}{\partial r}\Big|_{r=r_2} = -\kappa_3 \frac{\partial T_3}{\partial r}\Big|_{r=r_2} \\ T_3(r \rightarrow \infty) = -|\nabla T_0| r \cos\theta \end{cases} \tag{11.4}$$

where ∇T_0 represents the external thermal field.

For the convenience of comparison, throughout this chapter we set κ_3 to have the same value as the equivalent thermal conductivity κ_e of the core-shell structure. The temperature distribution in the host is given by Eq. (11.3). By applying the boundary conditions, we can derive the undetermined constants A_1 and C_1. Then we set C_1 to be zero to ensure the temperature distribution in the host undistorted. Namely, the ∇T_0 is maintained uniform without being affected by the core-shell structure. As a result, we obtain the expression for κ_e as

$$\kappa_e = \kappa_3 = c_2 \kappa_{rr_2} \frac{c_1 \kappa_{rr_1} + c_2 \kappa_{rr_2} + \left(c_1 \kappa_{rr_1} - c_2 \kappa_{rr_2}\right) p^{c_2}}{c_1 \kappa_{rr_1} + c_2 \kappa_{rr_2} - \left(c_1 \kappa_{rr_1} - c_2 \kappa_{rr_2}\right) p^{c_2}}. \tag{11.5}$$

Here $p = r_1^2/r_2^2$ is the area fraction of the core in the core-shell structure, and $c_i = \sqrt{\kappa_{\theta\theta_i}/\kappa_{rr_i}}$ with $i = 1$ or 2. In view of Eq. (11.5), if we set

$$c_1 \kappa_{rr_1} - c_2 \kappa_{rr_2} = 0, \tag{11.6}$$
$$c_1 \kappa_{rr_1} + c_2 \kappa_{rr_2} = 0, \tag{11.7}$$

they yield respectively

$$\kappa_e = c_1 \kappa_{rr_1} = c_2 \kappa_{rr_2}, \tag{11.8}$$
$$\kappa_e = c_1 \kappa_{rr_1} = -c_2 \kappa_{rr_2}. \tag{11.9}$$

Clearly, Eqs. (11.8) and (11.9) show that the resulting equivalent thermal conductivity κ_e is independent of the core's area fraction, thus being a constant. In other words, the area fraction can take any value within the full range from 0^+ to 1, which, however, does not affect the value of κ_e. More remarks can be added herein: in case of an isotropic core and shell, Eq. (11.9) predicts an equivalent thermal conductivity of the core-shell structure that equals the thermal conductivity of the core, which echoes with the condition required for the phenomenon of partial resonance in electromagnetic fields [19]. Here it is worth mentioning that Ref. [19] discuss only isotropic cases, which differ from anisotropic cases considered in this chapter.

Since thermal conductivities of the core and shell are coupled in Eqs. (11.6) and (11.7), we may call the two equations as two coupling relations. To distinguish the two relations, we name Eq. (11.6) [or Eq. (11.7)] as positive (or negative) relation. The physical meanings of both relations are to make the equivalent thermal conductivity of the core-shell structure independent of the area/volume fraction p.

11.3 Theory for Three Dimensions

The above theory for two dimensions (Fig. 11.1a) can be extended to three dimensions (Fig. 11.1b). Now κ_1 and κ_2 can be reformulated as $\kappa_1 = \mathrm{diag}\left(\kappa_{rr_1}, \kappa_{\theta\theta_1}, \kappa_{\varphi\varphi_1}\right)$ and $\kappa_2 = \mathrm{diag}\left(\kappa_{rr_2}, \kappa_{\theta\theta_2}, \kappa_{\varphi\varphi_2}\right)$ in spherical coordinates (r, θ, φ). For simplicity, we assume the system has an axial symmetry, namely, $\kappa_{\theta\theta} = \kappa_{\varphi\varphi}$. So, T is independent of φ. Then, the governing Laplace's equation can be written as

$$\frac{1}{r^2}\frac{\partial}{\partial r}\left(r^2\kappa_{rr}\frac{\partial T}{\partial r}\right) + \frac{1}{r\sin\theta}\frac{\partial}{\partial\theta}\left(\sin\theta\kappa_{\theta\theta}\frac{\partial T}{r\partial\theta}\right) = 0. \qquad (11.10)$$

Similar to the procedure for two dimensions, the equivalent thermal conductivity of the core-shell structure in three dimensions can be obtained as

$$\kappa_e = \kappa_3 = \kappa_{rr_2}$$
$$\times \frac{l_{21}\left(l_{11}\kappa_{rr_1} - l_{22}\kappa_{rr_2}\right) - l_{22}\left(l_{11}\kappa_{rr_1} - l_{21}\kappa_{rr_2}\right)p^{(l_{21}-l_{22})/3}}{\left(l_{11}\kappa_{rr_1} - l_{22}\kappa_{rr_2}\right) - \left(l_{11}\kappa_{rr_1} - l_{21}\kappa_{rr_2}\right)p^{(l_{21}-l_{22})/3}}, \qquad (11.11)$$

where $p = r_1^3/r_2^3$ is the volume fraction of the core in the core-shell structure, $l_{i1} = \left(-1 + \sqrt{1 + 8\kappa_{\theta\theta_i}/\kappa_{rr_i}}\right)/2$, and $l_{i2} = \left(-1 - \sqrt{1 + 8\kappa_{\theta\theta_i}/\kappa_{rr_i}}\right)/2$. Here $i = 1$ or 2.

Similarly, Eq. (11.11) yields two coupling relations

$$l_{11}\kappa_{rr_1} - l_{21}\kappa_{rr_2} = 0, \qquad (11.12)$$
$$l_{11}\kappa_{rr_1} - l_{22}\kappa_{rr_2} = 0, \qquad (11.13)$$

which, respectively, lead to

$$\kappa_e = l_{11}\kappa_{rr_1} = l_{21}\kappa_{rr_2}, \qquad (11.14)$$
$$\kappa_e = l_{11}\kappa_{rr_1} = l_{22}\kappa_{rr_2}. \qquad (11.15)$$

Clearly, the equivalent thermal conductivity κ_e in Eq. (11.14) [or Eq. (11.15)] is also independent of the core's volume fraction. Also, for clarity, we call Eq. (11.12) [or Eq. (11.13)] as positive (or negative) coupling relation.

11.4 Laboratory Experiments and Computer Simulations

In order to validate our theoretical analysis, now we are in a position to perform corresponding laboratory experiments and computer simulations.

In general, Eqs. (11.5) and (11.11) can predict the equivalent thermal conductivity of the core-shell structure in two or three dimensions. When the two coupling relations are satisfied [namely, Eqs. (11.6)–(11.7) for two dimensions and Eqs. (11.12)–(11.13) for three dimensions], the equivalent thermal conductivity of the core-shell structure will not change with the area or volume fraction. For comparison, we first utilize the commercial software COMSOL Multiphysics (https://www.comsol.com/) to perform finite-element simulations (Figs. 11.2 and 11.3a3–c3), which is followed by laboratory experiments (Fig. 11.3a1–c1 and 11.3a2–c2).

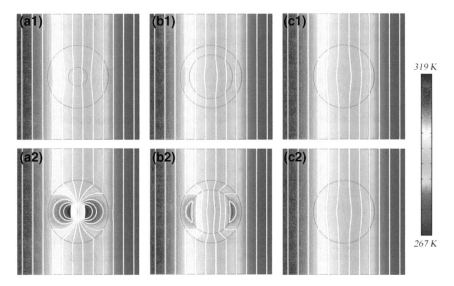

Fig. 11.2 Two-dimensional simulations for **a1–c1** positive relation [Eq. (11.6)] and **a2–c2** negative relation [Eq. (11.7)]. The color surface represents the temperature distribution, and the white lines represent the isotherms. The core and shell have anisotropic thermal conductivities and the host has an isotropic thermal conductivity. For reference, the pure core material is used in **c1** and **c2** instead, which has the same radius as the whole core-shell structure. Parameters: for **a1** and **a2**, $r_1 = 1.8$ and $r_2 = 6.0$ cm; for **b1** and **b2**, $r_1 = 4.2$ and $r_2 = 6.0$ cm; for **a1** and **b1**, $\kappa_1 =$ diag (194, 304) W/(m K), $\kappa_2 =$ diag (300, 197) W/(m K), and $\kappa_3 = 243$ W/(m K); for (**a2**) and (**b2**), $\kappa_1 =$ diag (194, 304) W/(m K), $\kappa_2 =$ diag $(-300, -197)$ W/(m K), and $\kappa_3 = 243$ W/(m K). Adapted from Ref. [20]

Figure 11.2 shows the finite-element simulations for two dimensions. We investigate the effects of both the positive relation [Eq. (11.6)] in Fig. 11.2a1, b1 and the negative relation [Eq. (11.7)] in Fig. 11.2a2, b2. Note that Fig. 11.2a2, b2 has a shell with apparently negative thermal conductivity. For reference, Fig. 11.2c1, c2 only has a core (without shell). All the hosts in the six panels Fig. 11.2a1–c2 are set to be the same, but the area fraction of the core is increased from Fig. 11.2a1 to c1 [or from Fig. 11.2a2 to c2]. Clearly, Fig. 11.2a1–c2 displays that the host's temperature distribution is identical to each other in the six panels. This behavior confirms the theoretical prediction that the equivalent thermal conductivity of the core-shell structure is independent of the core's area fraction under the two conditions described by Eq. (11.6) (positive relation) and Eq. (11.7) (negative relation).

Meanwhile, we perform experimental demonstration for the simulation results shown in Fig. 11.2a1–c1; see Fig. 11.3. Figure 11.3a1–c1 displays our three experimental samples, which are fabricated with copper by laser cutting. The left (or right) edges of the three samples are respectively put into the hot (or cold) sinks with constant temperatures. Then we use the FLIR E60 infrared camera to detect the temperature distribution of the samples, and the measurement results are shown in Fig. 11.3a2–c2, accordingly. By analyzing the temperature distribution in the host

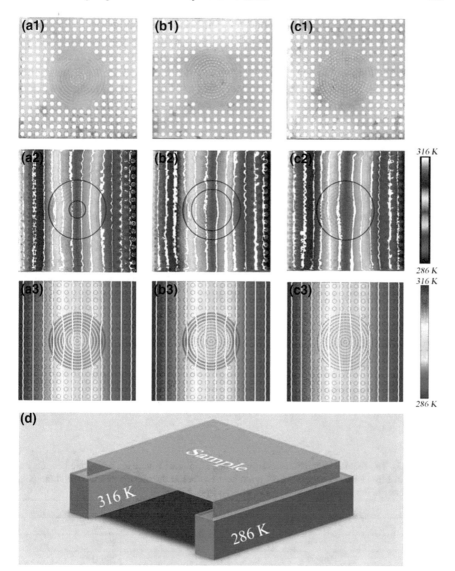

Fig. 11.3 Experimental demonstrations of Fig. 11.2a1–c1. **a1–c1** Are three experimental samples, each having a size of 24×24 cm and a thickness of 0.3 mm. The core-shell structure in **a1** or **b1** is composed of arrays made of ellipses with two different sizes: the large size is for the core, and the small size is for the shell. The core (or shell) is composed of red copper drilled with air ellipses, each having a major/minor semi-axis of 0.24/0.09 cm (or 0.18/0.045 cm). For reference, the pure core is used in (**c1**), which has the same radius as the core-shell structure shown in (**a1**) and (**b1**). The host in **a1–c1** is composed of red copper drilled with 204 air circular holes, each with radius 0.415 cm. In **a1–c1**, the thermal conductivities of the core and shell are diag (194, 304) W/(m K) and diag (300, 197) W/(m K). **a2–c2** [or **a3–c3**] are experimental measurements [or simulation results] of the three samples shown in **a1–c1**, respectively. Other parameters: thermal conductivity of red copper and air is 397 W/(m K) and 0.026 W/(m K), respectively; for **a1**, $r_1 = 1.8$ and $r_2 = 6.0$ cm; for **b1**, $r_1 = 4.2$ and $r_2 = 6.0$ cm. **d** Displays the experimental setup with a sample. Adapted from Ref. [20]

(see Fig. 11.3a2–c2), we confirm that the equivalent thermal conductivities of the core-shell structures shown in Fig. 11.3a1–c1 are (approximately) the same indeed. Furthermore, Fig. 11.3a3–c3 shows the computer simulations corresponding to the three samples shown in Fig. 11.3a1–c1, which just echoes with the experimental results (Fig. 11.3a2–c2) and the theoretical analysis (Fig. 11.2a1–c1). Figure 11.3d is an experimental setup.

As far as Fig. 11.2a2–c2 is concerned, we have adopted a shell with apparently negative thermal conductivity. Actually, one may resort to external energy to achieve apparently negative thermal conductivities [21–23]; see also Figs. 11.5 and 11.6 in the Supplementary Information. Apparently negative thermal conductivities and adding extra sources are equivalent only on the level of phenomena. In other words, apparently negative thermal conductivity can automatically generate a local high (or low) temperature which is impossible. To make it possible, we manually give a local high (or low) temperature to achieve the same phenomena without violating the second law of thermal dynamics. The shells with additional linear heat sources (Fig. 11.5a1, b1) or point heat sources (Fig. 11.5a2, b2) work the same effect as the shell with apparently negative thermal conductivity (Fig. 11.2a2, b2). The temperatures of the sources are presented in Tables 11.1, 11.2 and 11.3 in the Supplementary Information.

Besides, we perform finite-element simulations for three dimensions; see Fig. 11.4. Figure 11.4a1–c1 [or Fig. 11.4a2–c2] is simulation results based on Eq. (11.12) [or Eq. (11.13)]. Evidently, the temperature distribution outside the core-shell structure is also identical from Fig. 11.4a1 to c2, which agrees with the results shown in Fig. 11.2 for two dimensions.

Table 11.1 The required temperature value of each point heat source at the outside boundary of the shell in Fig. 11.5a2, b2: there are 24 point heat sources at the outside boundary and the point heat sources are numbered in a clockwise direction from 1 to 24

Source	Temp (K)	Source	Temp (K)
1	293.0	13	293.0
2	290.4	14	295.6
3	288.0	15	298.0
4	286.0	16	300.1
5	284.3	17	301.7
6	283.3	18	302.7
7	283.0	19	303.0
8	283.3	20	302.7
9	284.3	21	301.7
10	286.0	22	300.1
11	288.0	23	298.0
12	290.4	24	295.6

Table 11.2 The required temperature value of each point heat source at the inside boundary of the shell in Fig. 11.5a2: there are 16 point heat sources at the inside boundary and the point heat sources are numbered in a clockwise direction from 1 to 16

Source ($r_1 = 1.8$ cm)	Temp (K)
1	293.0
2	282.9
3	274.3
4	268.5
5	266.5
6	268.5
7	274.3
8	282.9
9	293.0
10	303.1
11	311.7
12	317.5
13	319.5
14	317.5
15	311.7
16	303.1

Table 11.3 The required temperature value of each point heat source at the inside boundary of the shell in Fig. 11.5b2: there are 16 point heat sources at the inside boundary and the point heat sources are numbered in a clockwise direction from 1 to 16

Source ($r_1 = 4.2$ cm)	Temp (K)
1	293.0
2	287.9
3	283.6
4	280.7
5	280.0
6	280.7
7	283.6
8	287.9
9	293.0
10	298.1
11	302.4
12	305.3
13	306.4
14	305.3
15	302.4
16	298.1

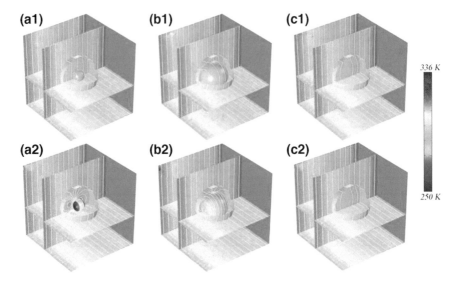

Fig. 11.4 Three-dimensional simulations for **a1**–**c1** positive relation [Eq. (11.12)] and **a2**–**c2** negative relation [Eq. (11.13)]. Others are the same as those in Fig. 11.2, but some parameters are changed: for **a1** and **b1**, $\kappa_1 = \mathrm{diag}\,(100, 417, 417)$ W/(m K) and $\kappa_2 = \mathrm{diag}\,(200, 269, 269)$ W/(m K); for **a2** and **b2**, $\kappa_1 = \mathrm{diag}\,(100, 417, 417)$ W/(m K) and $\kappa_2 = \mathrm{diag}\,(-200, -269, -269)$ W/(m K). Adapted from Ref. [20]

11.5 Discussion and Conclusion

This chapter proposed a scheme of correlated self-fixing behavior in thermal conduction. It has potential applications in thermal stabilization, for which one needs to overcome thermal fluctuations resulted from the changes of area/volume fractions due to thermal expansion or contraction, thermal stress concentration, etc.

Also, the uniform temperature distribution in the host of Figs. 11.2, 11.3 and 11.4 can help to hide the core-shell structure from being detected by infrared camera, which is useful for thermal camouflage or illusion [24–34].

Moreover, as aforementioned, the coupling mechanisms of Eq. (11.7) for two dimensions and Eq. (11.13) for three dimensions are similar to the partially resonant composites in electrostatics [19]. In Ref. [35], Milton et al. studied the cloaking effects associated with such partially resonant composites in electrostatics. Due to the similar dominant equation, related cloaking effects can be expected in thermotics as well.

In summary, we have solved Laplace's equation associated with appropriate boundary conditions, which helps to propose a class of intelligent thermal metamaterials based on a core-shell structure with anisotropic thermal conductivities. We have revealed two coupling mechanisms [i.e., Eqs. (11.6) and (11.7) for two dimensions or Eqs. (11.12) and (11.13) for three dimensions], which counter-intuitively cause the equivalent thermal conductivity of the core-shell structure to be always fixed at a constant value when the area or volume fraction of the core is changed within the full range (namely, from 0^+ to 1). Our theoretical results have been verified by both experiments and finite-element simulations. Nevertheless, our results are valid for steady states, rather than for unsteady states since we did not take into account the effects of mass density and heat capacity [36]. Besides thermotics, the present work also offers a different method on how to achieve similar intelligent metamaterials in other fields including electrostatics, magnetostatics and particle dynamics, which mathematically share the same dominant equation.

11.6 Supplementary Information

11.6.1 Approaches to Achieving Apparently Negative Thermal Conductivities: Computer Simulations

Since the negative coupling relations [Eqs. (11.7) and (11.13)] contain apparently negative thermal conductivities, we have to adopt external energy to obtain them, in order not to violate the second law of thermodynamics. By considering the uniqueness theorem, we can get the exact temperature values on the two boundaries of the shell. As a result, Fig. 11.5 shows that adding appropriate linear heat sources (Fig. 11.5a1, b1) or point heat sources (Fig. 11.5a2, b2) works the same as the shell with apparently negative thermal conductivity (Fig. 11.2a2, b2).

11.6.2 Approaches to Achieving Apparently Negative Thermal Conductivities: Laboratory Experiments

We further perform experiments to show the feasibility of applying external heat sources to achieve apparently negative thermal conductivities; see Fig. 11.6. Figure 11.6a is an experimental setup. In the experiment, the conductivity of the whole sample is made of copper with thermal conductivity 54 W/(m K), but we add external line heat sources to keep the temperature constant on the boundaries of $X = 4.5$ and $X = 17$ cm. Figure 11.6b is the temperature distribution of the sample shown in Fig. 11.6a, which is detected by using the FLIR E60 infrared camera. Figure 11.6c is the simulation result corresponding to the sample shown in Fig. 11.6a, which agrees with Fig. 11.6b. Figure 11.6d shows the simulation result of the temperature distri-

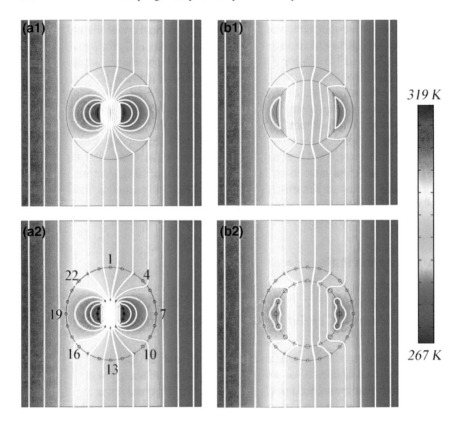

Fig. 11.5 Two-dimensional simulations for verifying Fig. 11.2a2, b2 [where apparently negative thermal conductivities have been adopted for the shell]. In **a1** and **b1**, we add external linear heat sources at the outside and inside boundaries of the shell, and the temperature at the outside (or inside) boundary can be determined by the function $293-50x$ (or $293-95.3827x$), where x is the horizontal axis of each point on the boundary. In **a2** and **b2**, we add external point heat sources at the outside and inside boundaries of the shell, and the temperatures at the boundaries are given in Tables 11.1, 11.2 and 11.3. Parameters: for **a1**, **a2**, $r_1 = 1.8$ and $r_2 = 6.0$ cm; for **b1**, **b2**, $r_1 = 4.2$ and $r_2 = 6.0$ cm; $\kappa_1 = \text{diag}(194, 304)$ W/(m K), $\kappa_2 = \text{diag}(300, 197)$ W/(m K), and $\kappa_3 = 243$ W/(m K). Adapted from Ref. [20]

bution, where the thermal conductivity of the middle material has a negative value, -40 W/(m K). Clearly, both Fig. 11.6c and d have the same temperature distribution. This behavior means that one can use additional energy to achieve apparently negative thermal conductivities indeed.

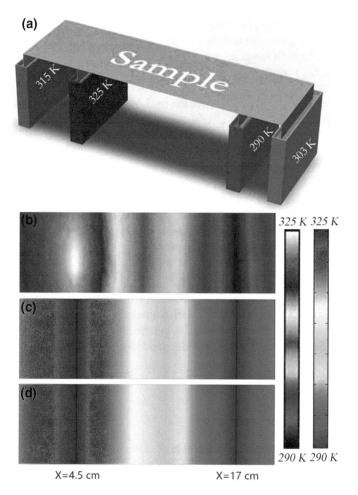

Fig. 11.6 Experimental demonstrations of apparently negative thermal conductivities. **a** Displays the experimental setup with a sample, whose size is 20×6 cm with thickness 0.3 mm. The thermal conductivity of the whole sample (made of copper) is 54 W/(m K). We apply two linear heat sources at the locations indicated by $X = 4.5$ and $X = 17$ cm. **b** Shows the temperature distribution of the sample shown in (**a**), which is experimentally detected by using the Flir E60 infrared camera. **c** Is simulation result of temperature distribution in the sample shown in (**a**). **d** Shows the simulation result of temperature distribution in the same sample (but without linear heat sources on the boundaries at $X = 4.5$ and $X = 17$ cm), whose middle region between $X = 4.5$ and $X = 17$ cm has been set to have a thermal conductivity with negative value, -40 W/(m K). Adapted from Ref. [20]

11.7 Exercises and Solutions

Exercises

1. Consider a two-dimensional case, a core-shell structure embedded in a host, whose center is located in the origin of coordinates; see Fig. 11.1a. The core (or shell)

has radius r_1 (or r_2) and thermal conductivity κ_1 (or κ_2). For the sake of generality, we assume the core and shell to be anisotropic. In polar coordinates (r, θ), κ_1 and κ_2 are both tensorial, which can be represented by $\mathrm{diag}\left(\kappa_{rr_1}, \kappa_{\theta\theta_1}\right)$ and $\mathrm{diag}\left(\kappa_{rr_2}, \kappa_{\theta\theta_2}\right)$, respectively. To focus on the core-shell structure, we simplify the host to be homogeneous and isotropic. In this case, for a passive and stable heat transport process, please derive the equivalent thermal conductivity κ_e of the core-shell structure and find out the condition under which κ_e is independent of the core's area fraction in the core-shell structure. $(\kappa_{\theta\theta_i}/\kappa_{rr_i} > 0, i = 1, 2)$.

2. The above Question 1 can be extended to three dimensions; see Fig. 11.1b. Now κ_1 and κ_2 can be reformulated as $\kappa_1 = \mathrm{diag}\left(\kappa_{rr_1}, \kappa_{\theta\theta_1}, \kappa_{\varphi\varphi_1}\right)$ and $\kappa_2 = \mathrm{diag}\left(\kappa_{rr_2}, \kappa_{\theta\theta_2}, \kappa_{\varphi\varphi_2}\right)$ in spherical coordinates (r, θ, φ). For simplicity, we assume the system has an axial symmetry, namely, $\kappa_{\theta\theta} = \kappa_{\varphi\varphi}$. So, T is independent of φ. Similarly, please derive the equivalent thermal conductivity κ_e of the core-shell structure and find out the condition under which κ_e is independent of the core's volume fraction in the core-shell structure.

Solutions

1. **Solution**: See Eq. (11.5) for the equivalent thermal conductivity and Eqs. (11.8) and (11.9) for the condition. Their derivations can be found in this chapter.

2. **Solution**: See Eq. (11.11) for the equivalent thermal conductivity and Eqs. (11.14) and (11.15) for the condition. Their derivations can be found in this chapter.

References

1. Maldovan, M.: Sound and heat revolutions in phononics. Nature **503**, 209–217 (2013)
2. Wegener, M.: Metamaterials beyond optics. Science **342**, 939–940 (2013)
3. Fan, C.Z., Gao, Y., Huang, J.P.: Shaped graded materials with an apparent negative thermal conductivity. Appl. Phys. Lett. **92**, 251907 (2008)
4. Chen, T.Y., Weng, C.N., Chen, J.S.: Cloak for curvilinearly anisotropic media in conduction. Appl. Phys. Lett. **93**, 114103 (2008)
5. Guenneau, S., Amra, C., Veynante, D.: Transformation thermodynamics: cloaking and concentrating heat flux. Opt. Express **20**, 8207–8218 (2012)
6. Narayana, S., Sato, Y.: Heat flux manipulation with engineered thermal materials. Phys. Rev. Lett. **108**, 214303 (2012)
7. Schittny, R., Kadic, M., Guenneau, S., Wegener, M.: Experiments on transformation thermodynamics: molding the flow of heat. Phys. Rev. Lett. **110**, 195901 (2013)
8. Hu, R., Xie, B., Hu, J.Y., Chen, Q., Luo, X.B.: Carpet thermal cloak realization based on the refraction law of heat flux. EPL (Europhys. Lett.) **111**, 54003 (2015)
9. Li, J.Y., Gao, Y., Huang, J.P.: A bifunctional cloak using transformation media. J. Appl. Phys. **108**, 074504 (2010)
10. Ma, Y.G., Liu, Y.C., Raza, M., Wang, Y.D., He, S.L.: Experimental demonstration of a multiphysics cloak: manipulating heat flux and electric current simultaneously. Phys. Rev. Lett. **113**, 205501 (2014)
11. Stedman, T., Woods, L.M.: Cloaking of thermoelectric transport. Sci. Rep. **7**, 6988 (2017)
12. Moccia, M., Castaldi, G., Savo, S., Sato, Y., Galdi, V.: Independent manipulation of heat and electrical current via bifunctional metamaterials. Phys. Rev. X **4**, 021025 (2014)

13. Lan, C.W., Li, B., Zhou, J.: Simultaneously concentrated electric and thermal fields using fan-shaped structure. Opt. Express **23**, 24475 (2015)
14. Peng, R.G., Xiao, Z.Q., Zhao, Q., Zhang, F.L., Meng, Y.G., Li, B., Zhou, J., Fan, Y.C., Zhang, P., Shen, N.-H., Koschny, T., Soukoulis, C.M.: Temperature-controlled chameleonlike cloak. Phys. Rev. X **7**, 011033 (2017)
15. Xu, L.J., Yang, S., Huang, J.P.: Thermal theory for heterogeneously architected structure: fundamentals and application. Phys. Rev. E **98**, 052128 (2018)
16. Li, Y., Shen, X.Y., Wu, Z.H., Huang, J.Y., Chen, Y.X., Ni, Y.S., Huang, J.P.: Temperature-dependent transformation thermotics: from switchable thermal cloaks to macroscopic thermal diodes. Phys. Rev. Lett. **115**, 195503 (2015)
17. Shen, X.Y., Li, Y., Jiang, C.R., Ni, Y.S., Huang, J.P.: Thermal cloak-concentrator. Appl. Phys. Lett. **109**, 031907 (2016)
18. Carslaw, H.S., Jaeger, J.C.: Conduction of Heat in Solids, 2nd edn. Clarendon Press, Oxford (1959)
19. Nicorovici, N.A., McPhedran, R.C., Milton, G.W.: Optical and dielectric properties of partially resonant composites. Phys. Rev. B **49**, 8479 (1994)
20. Yang, S., Xu, L.J., Huang, J.P.: Intelligence thermotics: correlated self-fixing behavior of thermal metamaterials. EPL (Europhys. Lett.) **126**, 54001 (2019)
21. Gao, Y., Huang, J.P.: Unconventional thermal cloak hiding an object outside the cloak. EPL (Europhys. Lett.) **104**, 44001 (2013)
22. Nguyen, D.M., Xu, H.Y., Zhang, Y.M., Zhang, B.L.: Active thermal cloak. Appl. Phys. Lett. **107**, 121901 (2015)
23. Lan, C.W., Bi, K., Gao, Z.H., Li, B., Zhou, J.: Achieving bifunctional cloak via combination of passive and active schemes. Appl. Phys. Lett. **109**, 201903 (2016)
24. Han, T.C., Bai, X., Thong, J.T.L., Li, B.W., Qiu, C.-W.: Full control and manipulation of heat signatures: cloaking, camouflage and thermal metamaterials. Adv. Mat. **26**, 1731–1734 (2014)
25. He, X., Wu, L.Z.: Illusion thermodynamics: a camouflage technique changing an object into another one with arbitrary cross section. Appl. Phys. Lett. **105**, 221904 (2014)
26. Yang, T.Z., Bai, X., Gao, D.L., Wu, L.Z., Li, B.W., Thong, J.T.L., Qiu, C.W.: Invisible sensors: simultaneous sensing and camouflaging in multiphysical fields. Adv. Mater. **27**, 7752–7758 (2015)
27. Yang, T.Z., Su, Y.S., Xu, W.K., Yang, X.D.: Transient thermal camouflage and heat signature control. Appl. Phys. Lett. **109**, 121905 (2016)
28. Xu, L.J., Jiang, C.R., Shang, J., Wang, R.Z., Huang, J.P.: Periodic composites: quasi-uniform heat conduction, Janus thermal illusion, and illusion thermal diodes. Eur. Phys. J. B **90**, 221 (2017)
29. Hu, R., Zhou, S.L., Li, Y., Lei, D.Y., Luo, X.B., Qiu, C.W.: Illusion thermotics. Adv. Mater. **30**, 1707237 (2018)
30. Zhou, S.L., Hu, R., Luo, X.B.: Thermal illusion with twinborn-like heat signatures. Int. J. Heat Mass Transf. **127**, 607 (2018)
31. Xu, L.J., Wang, R.Z., Huang, J.P.: Camouflage thermotics: a cavity without disturbing heat signatures outside. J. Appl. Phys. **123**, 245111 (2018)
32. Xu, L.J., Huang, J.P.: A transformation theory for camouflaging arbitrary heat sources. Phys. Lett. A **382**, 3313 (2018)
33. Li, Y., Bai, X., Yang, T.Z., Luo, H., Qiu, C.W.: Structured thermal surface for radiative camouflage. Nat. Commun. **9**, 273 (2018)
34. Wang, R.Z., Xu, L.J., Ji, Q., Huang, J.P.: A thermal theory for unifying and designing transparency, concentrating and cloaking. J. Appl. Phys. **123**, 115117 (2018)
35. Milton, G.W., Nicorovici, N.A.P.: On the cloaking effects associated with anomalous localized. Proc. R. Soc. A **462**, 3027–3059 (2006)
36. Sklan, S.R., Bai, X., Li, B.W., Zhang, X.: Detecting thermal cloaks via transient effects. Sci. Rep. **6**, 32915 (2016)

Chapter 12
Coupling Theory for Temperature-Dependent Thermal Conductivities: Nonlinearity Modulation and Enhancement

Abstract Thermal metamaterials based on core-shell structures have aroused wide research interest, e.g., thermal cloaks. However, almost all the relevant studies only discuss linear materials whose thermal conductivities are temperature-independent constants. Nonlinear materials (whose thermal conductivities depend on temperatures) have seldom been touched, which, however, are important in practical applications. This situation largely results from the lack of a theoretical framework for handling such nonlinear problems. Here we study the nonlinear responses of thermal metamaterials with a core-shell structure in two or three dimensions. By calculating the effective thermal conductivity, we derive the nonlinear modulation of a nonlinear core. Furthermore, we reveal two thermal coupling conditions, under which this nonlinear modulation can be efficiently manipulated. In particular, we reveal the phenomenon of nonlinearity enhancement. Then this theory helps us to design a kind of intelligent thermal transparency devices, which can respond to the direction of thermal fields. The theoretical results and finite-element simulations agree well with each other. This chapter not only offers a different mechanism to achieve nonlinearity modulation and enhancement in thermotics, but also suggests potential applications in thermal management including illusion.

Keywords Nonlinearity modulation · Nonlinearity enhancement · Core-shell structure · Temperature-dependent thermal conductivity · Coupling condition · Intelligent thermal transparency

12.1 Opening Remarks

Heat management has aroused intensive research interest due to its wide applications for human beings. One core problem of heat management is to tailor thermal conductivities effectively. Fortunately, thermal metamaterials have provided a powerful method to tailor thermal conductivities with delicately designed structures. Based on thermal metamaterials, a large amount of novel thermal phenomena have been realized, such as thermal cloaks [1–8], thermal concentrators [3, 4], thermal rotators [9, 10], thermal transparency [11–16], thermal camouflage [17–23], thermal

© Springer Nature Singapore Pte Ltd. 2020
J.-P. Huang, *Theoretical Thermotics*,
https://doi.org/10.1007/978-981-15-2301-4_12

bending [24–27], etc. For achieving these phenomena, the core-shell structure serve as a typical scheme. However, the existing research does not consider the nonlinear effect (nonlinear thermotics) except for some piecemeal studies [28–32]. Compared with nonlinear optics [33–38], nonlinear thermotics has attracted much less attention. This situation largely results from the lack of a general theoretical framework to handle nonlinear effects in thermotics.

To solve this problem, here we investigate the thermal properties of a core-shell structure embedded in a finite matrix. The core is nonlinear, and the shell and the matrix are linear. Here, the nonlinear core (or linear shell/matrix) means that the core (or shell/matrix) material has a temperature-dependent (or temperature-independent) thermal conductivity; in this case, corresponding Fourier's law of thermal conduction shows a nonlinear (or linear) relation between the heat flux density and the temperature gradient, thus called "nonlinear core" (or "linear shell/matrix"). Then we establish a general theoretical framework to deal with nonlinear effects in both two and three dimensions. To achieve nonlinearity enhancement, we discuss the nonlinear modulation under two thermal coupling conditions after establishing the general theory. This is because under thermal coupling conditions, the core property can be extended to the shell. In this way, the core nonlinearity may also be extended to the shell, which is beneficial for our purpose. Moreover, thermal coupling conditions largely simplify the mathematical form of the nonlinear modulation. Results indicate that the nonlinearity enhancement can appear under one of the coupling conditions. Further, the theory helps us to propose a kind of intelligent thermal transparency devices, which become automatically switchable to external temperatures. We also perform finite-element simulations to validate our theoretical predictions, and they agree well with each other.

12.2 Theory

12.2.1 Two-Dimensional Case

We first consider the two-dimensional case; see Fig. 12.1a. The core-shell structure is embedded in a finite square matrix with width a and temperature-independent (namely, linear) thermal conductivity κ_m. The shell with radius r_2 has an anisotropic linear thermal conductivity $\bar{\kappa}_s = \mathrm{diag}\,(\kappa_{rr},\ \kappa_{\theta\theta})$ in cylindrical coordinates $(r,\ \theta)$. The core with radius r_1 owns a temperature-dependent (i.e., nonlinear) thermal conductivity $\kappa_c\,(T)$ given by [32]

$$\kappa_c\,(T) = \kappa_c^{(0)} + \chi_c T^\alpha, \tag{12.1}$$

where $\kappa_c^{(0)}$ is the temperature-independent (or linear) part, χ_c and T, respectively, represent nonlinear coefficient and temperature, and α can be any real number. We assume that the core is weakly temperature-dependent (or nonlinear), say

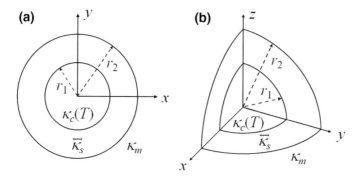

Fig. 12.1 Schematic diagrams of **a** two-dimensional or **b** three-dimensional core-shell structure in Cartesian coordinate system; **b** only shows one eighth of the structure for the sake of clarity. The core or shell radius is r_1 or r_2; the core, shell and matrix have a thermal conductivity of $\kappa_c(T)$, $\bar{\kappa}_s$, and κ_m, respectively. Other details can be found in the text. Adapted from Ref. [46]

$\chi_c T^\alpha \ll \kappa_c^{(0)}$. Regarding of Eq. (12.1), it becomes necessary for us to add two remarks as follows. On one hand, we need to point out that in nonlinear optics, dielectric permittivities have a similar nonlinear expression because of their dependence on the electric-field component of electromagnetic waves. But, this electric field is mathematically analogous to the temperature gradient in thermotics, rather than the temperature (T) as adopted in Eq. (12.1). This fact implies that new physics may be expected from Eq. (12.1). On the other hand, Eq. (12.1) has realistic implications because almost all materials have temperature-dependent thermal conductivities (certainly, the dependence could be weak or strong). Particularly, thermal conductivity κ_c could increase as T either increases [39] or decreases [40].

Then, the effective thermal conductivity of the core-shell structure $\kappa_{e1}(T)$ [15] reaches

$$\kappa_{e1}(T) = u\kappa_{rr} \frac{\kappa_c(T) + u\kappa_{rr} + [\kappa_c(T) - u\kappa_{rr}]\, p_1^u}{\kappa_c(T) + u\kappa_{rr} - [\kappa_c(T) - u\kappa_{rr}]\, p_1^u}, \tag{12.2}$$

and that of the core-shell structure plus the matrix $\kappa_{e2}(T)$ [14] turns to be

$$\kappa_{e2}(T) = \kappa_m \frac{\kappa_{e1}(T) + \kappa_m + [\kappa_{e1}(T) - \kappa_m]\, p_2}{\kappa_{e1}(T) + \kappa_m - [\kappa_{e1}(T) - \kappa_m]\, p_2}, \tag{12.3}$$

where $p_1 = r_1^2/r_2^2$, $p_2 = \pi r_2^2/a^2$ and $u = \sqrt{\kappa_{\theta\theta}/\kappa_{rr}}$. The direct use of the results from Refs. [14] and [15] is valid in the paper. The reasons lie in that (i) the nonlinear term is smaller than the linear term, and thus the spatial fluctuations of the thermal conductivity are small; (ii) we assume that T is the temperature at the center of the structure. So the assumptions can average over the spatial fluctuations of the thermal conductivity. Therefore, the results from Refs. [14] and [15] are approximately valid and still contributing.

Equation (12.3) allows us to do Taylor expansion up to infinite terms. In the following, we keep the terms up to $T^{3\alpha}$, and neglect the other terms,

$$\kappa_{e2}(T) = \kappa_{e2}^{(0)} + \chi_e T^\alpha + \beta_e T^{2\alpha} + \gamma_e T^{3\alpha} + O\left(T^{4\alpha}\right), \tag{12.4}$$

where $\kappa_{e2}^{(0)}$, χ_e, β_e, and γ_e are respectively

$$\kappa_{e2}^{(0)} = \kappa_m \frac{\kappa_{e1}^{(0)} + \kappa_m + \left(\kappa_{e1}^{(0)} - \kappa_m\right) p_2}{\kappa_{e1}^{(0)} + \kappa_m - \left(\kappa_{e1}^{(0)} - \kappa_m\right) p_2}, \tag{12.5}$$

$$\chi_e = \frac{16 u^2 \kappa_{rr}^2 \kappa_m^2 p_2 p_1^u \chi_c}{\left\{u\kappa_{rr}(p_2 - 1)\left[\kappa_c^{(0)} + u\kappa_{rr} + (\kappa_c^{(0)} - u\kappa_{rr})p_1{}^u\right] + \kappa_m(p_2 + 1)\left[\kappa_c^{(0)} + u\kappa_{rr} - (\kappa_c^{(0)} - u\kappa_{rr})p_1{}^u\right]\right\}^2}, \tag{12.6}$$

$$\beta_e = \frac{16 u^2 \kappa_{rr}^2 \kappa_m^2 p_2 p_1^u \chi_c^2 \left[u\kappa_{rr}\left(p_1^u + 1\right)(p_2 - 1) + \kappa_m\left(p_1^u - 1\right)(p_2 + 1)\right]}{\left\{u\kappa_{rr}(p_2 - 1)\left[\kappa_c^{(0)} + u\kappa_{rr} + (\kappa_c^{(0)} - u\kappa_{rr})p_1{}^u\right] + \kappa_m(p_2 + 1)\left[\kappa_c^{(0)} + u\kappa_{rr} - (\kappa_c^{(0)} - u\kappa_{rr})p_1{}^u\right]\right\}^3}, \tag{12.7}$$

$$\gamma_e = \frac{16 u^2 \kappa_{rr}^2 \kappa_m^2 p_2 p_1^u \chi_c^3 \left[u\kappa_{rr}\left(p_1^u + 1\right)(p_2 - 1) + \kappa_m\left(p_1^u - 1\right)(p_2 + 1)\right]^2}{\left\{u\kappa_{rr}(p_2 - 1)\left[\kappa_c^{(0)} + u\kappa_{rr} + (\kappa_c^{(0)} - u\kappa_{rr})p_1{}^u\right] + \kappa_m(p_2 + 1)\left[\kappa_c^{(0)} + u\kappa_{rr} - (\kappa_c^{(0)} - u\kappa_{rr})p_1{}^u\right]\right\}^4}. \tag{12.8}$$

Here $\kappa_{e2}^{(0)}$ is the linear part of the thermal conductivity of the core-shell structure and the matrix, and $\kappa_{e1}^{(0)} = u\kappa_{rr}\left[\kappa_c^{(0)} + u\kappa_{rr} + \left(\kappa_c^{(0)} - u\kappa_{rr}\right) p_1^u\right] / \left[\kappa_c^{(0)} + u\kappa_{rr} - \left(\kappa_c^{(0)} - u\kappa_{rr}\right) p_1^u\right]$ is the the linear part of the thermal conductivity of the core-shell structure.

Equation (12.4) clearly shows that the low-order nonlinearity (Eq. (12.1)) can induce not only the same order nonlinearity $\chi_e T^\alpha$, but also the high-order nonlinearities (i.e., $\beta_e T^{2\alpha}$ and $\gamma_e T^{3\alpha}$). Nevertheless, owing to $\chi_c T^\alpha \ll \kappa_c^{(0)}$ in Eq. (12.1), it is evident to conclude that $\chi_e T^\alpha \gg \beta_e T^{2\alpha} \gg \gamma_e T^{3\alpha}$ in Eq. (12.4). So, in what follows, we only focus on $\chi_e T^\alpha$. To proceed, we define the nonlinear modulation $\eta = \chi_e/\chi_c$, which is given by

$$\eta = \frac{16 u^2 \kappa_{rr}^2 \kappa_m^2 p_2 p_1^u}{\left\{u\kappa_{rr}(p_2 - 1)\left[\kappa_c^{(0)} + u\kappa_{rr} + (\kappa_c^{(0)} - u\kappa_{rr})p_1{}^u\right] + \kappa_m(p_2 + 1)\left[\kappa_c^{(0)} + u\kappa_{rr} - (\kappa_c^{(0)} - u\kappa_{rr})p_1{}^u\right]\right\}^2}. \tag{12.9}$$

This equation is the general expression of nonlinear modulation in two dimensions. Then we are allowed to discuss the nonlinear modulation η in some special cases, say, under thermal coupling conditions. Namely, when the core-shell structure satisfies

$$\kappa_c^{(0)} + u\kappa_{rr} = 0, \tag{12.10}$$

$$\kappa_m = \kappa_c^{(0)}, \tag{12.11}$$

the nonlinear modulation η is simplified as

$$\eta = p_1^{-u} p_2. \tag{12.12}$$

On the other hand, when the core-shell structure satisfies

$$\kappa_c^{(0)} - u\kappa_{rr} = 0, \tag{12.13}$$

$$\kappa_m = \kappa_c^{(0)}, \tag{12.14}$$

the nonlinear modulation η becomes

$$\eta = p_1^u p_2. \tag{12.15}$$

Equations (12.10, 12.11) and (12.13, 12.14) are two different thermal coupling conditions because they establish the relations among the core, shell, and matrix.

12.2.2 Three-Dimensional Case

The above theory can be extended to three dimensions; see Fig. 12.1b. Then $\bar{\kappa}_s$ should be redefined as $\bar{\kappa}_s = \mathrm{diag}\left(\kappa_{rr}, \kappa_{\theta\theta}, \kappa_{\varphi\varphi}\right)$ in spherical coordinates (r, θ, φ) with $\kappa_{\theta\theta} = \kappa_{\varphi\varphi}$ for simplicity. The effective thermal conductivity of the core-shell structure [15] in three dimensions can be expressed as

$$\kappa_{e1}(T) = \kappa_{rr} \frac{v_1\left[\kappa_c(T) - v_2\kappa_{rr}\right] - v_2\left[\kappa_c(T) - v_1\kappa_{rr}\right]p_1^w}{\left[\kappa_c(T) - v_2\kappa_{rr}\right] - \left[\kappa_c(T) - v_1\kappa_{rr}\right]p_1^w}, \tag{12.16}$$

and that of the core-shell structure plus the matrix [14] is,

$$\kappa_{e2}(T) = \kappa_m \frac{\kappa_{e1}(T) + 2\kappa_m + 2\left[\kappa_{e1}(T) - \kappa_m\right]p_2}{\kappa_{e1}(T) + 2\kappa_m - \left[\kappa_{e1}(T) - \kappa_m\right]p_2}, \tag{12.17}$$

where $p_1 = (r_1/r_2)^3$, $p_2 = 4\pi r_2^3/3a^3$, $v_{1,2} = -1/2 \pm \sqrt{1/4 + 2\kappa_{\theta\theta}/\kappa_{rr}}$, and $w = \sqrt{1 + 8\kappa_{\theta\theta}/\kappa_{rr}}/3$.

Similar to the procedure in two dimensions, the nonlinear modulation is

$$\eta = \frac{81\kappa_m^2\kappa_{rr}^2 w^2 p_2 p_1^w}{\left\{\kappa_{rr}(1 - p_2)\left[v_1(\kappa_c^{(0)} - v_2\kappa_{rr}) - v_2(\kappa_c^{(0)} - v_1\kappa_{rr})p_1^w\right] + \kappa_m(2 + p_2)\left[(\kappa_c^{(0)} - v_2\kappa_{rr}) - (\kappa_c^{(0)} - v_1\kappa_{rr})p_1^w\right]\right\}^2}. \tag{12.18}$$

Equation (12.18) is the general expression of nonlinear modulation in three dimensions. Then we will also discuss the nonlinear modulation under thermal coupling conditions.

When the core-shell structure satisfies

$$\kappa_c^{(0)} - v_2\kappa_{rr} = 0, \tag{12.19}$$

$$\kappa_m = \kappa_c^{(0)}, \tag{12.20}$$

the nonlinear modulation turns to be

$$\eta = p_1^{-w} p_2. \tag{12.21}$$

On the other hand, when the core-shell structure meets

$$\kappa_c^{(0)} - v_1 \kappa_{rr} = 0, \tag{12.22}$$

$$\kappa_m = \kappa_c^{(0)}, \tag{12.23}$$

the nonlinear modulation reaches

$$\eta = p_1^{w} p_2. \tag{12.24}$$

Alike, Eqs. (12.19, 12.20) and (12.22, 12.23) are also two different thermal coupling conditions in three dimensions.

12.3 Theoretical Calculation Versus Finite-Element Simulation

We have established a theoretical framework to handle the nonlinear modulation (η) in both two and three dimensions, especially under thermal coupling conditions. Now we are in a position to validate the predicted η with finite-element simulations. For this realization, we first calculate the effective thermal conductivity $\kappa_{e2}(T)$ with $J/|\nabla T_0|$, where J is the overall average heat flux obtained from COMSOL Multiphysics (http://www.comsol.com/). In Eq. (12.4), $\kappa_{e2}^{(0)}$ can be theoretically calculated with Eq. (12.5). T can be approximately regarded as the temperature at the center since the nonlinearity of the system is not that strong. In this way, we can derive η with $\left[\kappa_{e2}(T) - \kappa_{e2}^{(0)} \right] / (\chi_c T^\alpha)$ based on finite-element simulations; see symbols in Figs. 12.2 and 12.3. Then we explore the nonlinear modulation η when thermal coupling conditions are satisfied, namely, Eqs. (12.12, 12.15) for two dimensions and Eqs. (12.21, 12.24) for three dimensions in theory; see lines in Figs. 12.2 and 12.3.

First we analyze the two-dimensional case whose results are presented in Fig. 12.2. Figure 12.2a1–c1 shows the nonlinear modulation under the thermal coupling condition determined by Eqs. (12.10, 12.11). According to the theoretical analysis of Eq. (12.12), the nonlinear modulation is related to three key parameters, say, the degree of shell anisotropy $\kappa_{\theta\theta}/\kappa_{rr}$, the core fraction in the shell p_1, and the core-shell fraction in the matrix p_2. It is noted that the maximum value of p_2 is $\pi/4$ because a circle cannot fill up a square. The nonlinear modulation η can be well manipulated and enhanced under the thermal coupling condition determined by Eqs. (12.10, 12.11). Here the word "enhanced" means $\eta > 1$, which indicates that χ_e (effective nonlinear coefficient) is counter-intuitively larger than χ_c (core's nonlinear coefficient). However, the nonlinear modulation η cannot be enhanced (namely, η is always smaller

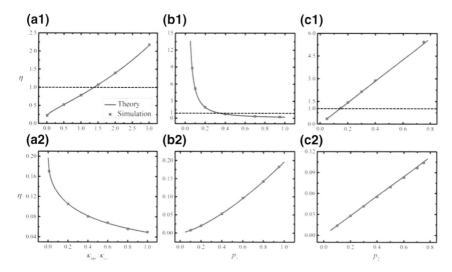

Fig. 12.2 Two-dimensional simulation results. **a1–c1** Show the nonlinear modulation η ($=\chi_e/\chi_c$) under the thermal coupling conditions determined by Eqs. (12.10, 12.11) with three variables: **a1** $\kappa_{\theta\theta}/\kappa_{rr}$, **b1** p_1, and **c1** p_2. The solid lines in **a1–c1** are calculated from Eq. (12.12), and the symbols are obtained from finite-element simulations. The solid lines in **a2–c2** are calculated from Eq. (12.15) under the thermal coupling conditions determined by Eqs. (12.13, 12.14). Other parameters: **a1, a2** $p_1 = 0.25$ and $p_2 = \pi/16$; **b1, b2** $\kappa_{\theta\theta}/\kappa_{rr} = 2$ and $p_2 = \pi/16$; **c1, c2** $\kappa_{\theta\theta}/\kappa_{rr} = 2$ and $p_1 = 0.25$; **a1–c1** $\kappa_c(T) = 400 + 0.05T$ W/(m K), $\kappa_s = -\sqrt{\kappa_{rr}\kappa_{\theta\theta}} = -400$ W/(m K) is the effective scalar thermal conductivity, and $\kappa_m = 400$ W/(m K); **a2–c2** $\kappa_c(T) = 400 + 0.05T$ W/(m K), $\kappa_s = \sqrt{\kappa_{rr}\kappa_{\theta\theta}} = 400$ W/(m K), and $\kappa_m = 400$ W/(m K). Adapted from Ref. [46]

than 1) under the thermal coupling condition determined by Eqs. (12.13, 12.14), no matter how one adjusts the three associated parameters.

Then we discuss the three-dimensional case whose results are displayed in Fig. 12.3. Figure 12.3a1–c1 (or Fig. 12.3a2–c2) displays the thermal coupling condition determined by Eqs. (12.19, 12.20) (or Eqs. (12.22, 12.23)). Similar conclusion can be obtained. That is, only the thermal coupling condition determined by Eqs. (12.19, 12.20) succeeds in achieving nonlinearity enhancement (i.e., $\eta > 1$), whereas the thermal coupling condition determined by Eqs. (12.22, 12.23) fails.

As shown in Figs. 12.2 and 12.3, the finite-element simulation results agree well with the theoretical calculations, and the nonlinearity can be enhanced up to one order of magnitude when the physical parameters are chosen appropriately.

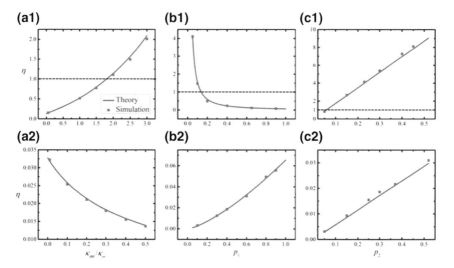

Fig. 12.3 Three-dimensional simulation results. The solid lines in **a1–c1** (or (**a2–c2**)) are calculated from Eq. (12.21) (or Eq. (12.24)), and the symbols are obtained from finite-element simulations. Other parameters are the same as those in Fig. 12.2 except for (**a1, a2**) $p_1 = 0.125$ and $p_2 = \pi/48$, **b1, b2** $\kappa_{\theta\theta}/\kappa_{rr} = 2$ and $p_2 = \pi/48$, and **c1, c2** $\kappa_{\theta\theta}/\kappa_{rr} = 2$ and $p_1 = 0.125$. Adapted from Ref. [46]

12.4   Application of Nonlinearity

In addition, based on the proposed theory, here we design an intelligent (switchable) thermal transparency device; see Fig. 12.4.

Traditional thermal transparency can ensure the external thermal fields undistorted [11–16]. However, it is independent of the direction of the thermal fields, which may lack the intelligence for controllability between "open" and "close" state. Here the nonlinear property helps to control the thermal transparency with respect to different directions of the thermal fields, thus being called intelligent thermal transparency.

In Fig. 12.4a–c, the device has a nonlinear core and a linear shell. To achieve the effect of switching, here we split the core into two parts. Two kinds of nonlinear thermal conductivities can respond to different boundary conditions (the direction of heat flux) automatically. When the heat flux goes from left to right (Fig. 12.4b), the temperature is uniformly distributed in the matrix, thus yielding the phenomenon of thermal transparency. In this case, the device is on "open" state; see Fig. 12.4b. Conversely, if the heat flux moves from right to left (Fig. 12.4c), the core-shell structure will affect the temperature distribution of matrix, thus eliminating the behavior of thermal transparency. As a result, the device is on "close" state; see Fig. 12.4c. Namely, the switching function of the thermal transparency device is achieved as expected.

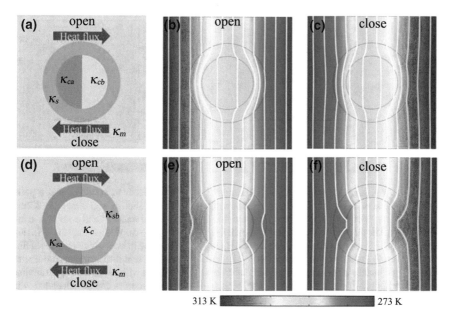

Fig. 12.4 Switchable thermal transparency device: **a** is the schematic diagram based on the nonlinear core and linear shell and **b, c** are corresponding finite-element simulation results. **b** is on open state and **c** is on close state. **d** is the schematic diagram based on the linear core and nonlinear shell and **e, f** are corresponding finite-element simulation results. **e** is on open state and **f** is on close state. Other parameters in **a–f**: $\kappa_{ca} = 400 + 70 \times (T - 293)$ W/(m K), $\kappa_{cb} = 400 - 70 \times (T - 293)$ W/(m K), $\kappa_s = 100$ W/(m K), $\kappa_c = 20$ W/(m K), $\kappa_{sa} = 400 + 8 \times (T - 293)$ W/(m K), $\kappa_{sb} = 400 - 8 \times (T - 293)$ W/(m K), and $\kappa_m = 200$ W/(m K). Here both κ_{ca}, κ_{cb}, κ_{sa} and κ_{sb} are temperature-dependent, which could, in principle, be designed by using shape memory alloys according to the method proposed in Refs. [28, 31]. Adapted from Ref. [46]

Also, in Fig. 12.4d–f, we design a nonlinear shell and a linear core. The similar results can also be achieved; see Fig. 12.4e, f. A comment on Fig. 12.4a–c and d–f is that in spite of the similar switching phenomena, the nonlinearity of the shell (Fig. 12.4d) can be smaller than that of the core (Fig. 12.4a), which means that the manipulation of the shell nonlinearity is more efficient.

12.5 Discussion and Conclusion

Nonlinearity (namely, thermally-responsive thermal conductivity) is of great significance to achieve thermal management. Although natural materials such as copper may exhibit weak nonlinearity, they are still not strong enough to achieve practical nonlinear effects in certain situations. In this chapter, we have investigated the nonlinear modulation of a core-shell structure embedded in a finite matrix (only the core is nonlinear). Under two thermal coupling conditions, the nonlinear modulation can be largely simplified, and only depends on three key parameters: the

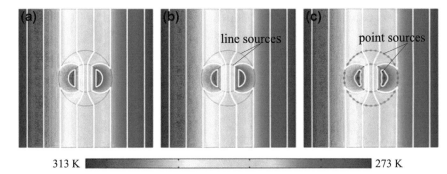

Fig. 12.5 Replacing **a** apparent negative thermal conductivity with **b** line sources and **c** point sources. The simulation box of **a–c** is 16×16 cm. The radius of shell r_s and core r_c is 3.2 cm and 1.6 cm, respectively. The thermal conductivities of background material and core in **a–c** are 50 W/(m K). The thermal conductivities of the shells in **a** and **b**, **c** are -50 W/(m K) and 20 W/(m K), respectively. Line sources are applied on the two boundaries of the shell in **b**. The temperatures of line sources obey $T = 293 - 250 r_{c,s}^2 x / \left(x^2 + y^2\right)$ K, where (x, y) represent the Cartesian coordinates whose origin locates in the center of the simulation box. Point sources with radius 0.1 cm are applied on the two boundaries of the shell in **c**. The temperatures of point sources can be calculated with $T = 293 - 250 r_{c,s}^2 x / \left(x^2 + y^2\right)$ K according to the source positions, respectively. Adapted from Ref. [46]

degree of shell anisotropy, the core fraction in the shell, and the core-shell fraction in the matrix. Therefore, we can achieve the aim of regulating the nonlinearity by the three tunable parameters. In particular, the nonlinear modulation will be effectively enhanced under the thermal coupling conditions determined by Eqs. (12.10, 12.11) and Eqs. (12.19, 12.20). Our work lays the foundation for studying the nonlinear property of a core-shell structure, and further work can be expected to explore more complicated cases like nonlinear shells or nonlinear matrices.

In the process to achieve nonlinearity enhancement, apparent negative thermal conductivities [41–44] are applied, which means that the direction of heat flux is from low temperature to high temperature. For this realization, a reliable way is to add external energy to avoid violating the second law of thermodynamics. We also perform finite-element simulations to verify the feasibility of apparent negative thermal conductivities; see Fig. 12.5. We add external line sources (Fig. 12.5b) and point sources (Fig. 12.5c) on the two boundaries of the shell, and set the thermal conductivity of the shell to be positive. The temperature distributions in Fig. 12.5b, c are the same as that in Fig. 12.5a. Therefore, it is contributing to add external energy. Such point sources can be realized by experiment; see the experimental setup shown in Fig. 1 of Ref. [45].

In summary, this chapter extends nonlinearity research from optics to thermotics, but with essential difference in the definition of nonlinearity (Eq. (12.1)). We have proposed a different mechanism to modulate nonlinear thermal responses, and achieved both thermal nonlinearity enhancement and intelligent thermal transparency under various kinds of conditions. We expect that the nonlinearity studied in this chapter could also have potential applications in heat management including illusion.

12.6   Exercises and Solutions

Exercises

1. Consider a two-dimensional case; see Fig. 12.1. The core-shell structure is embedded in a finite square matrix with width a and temperature-independent (namely, linear) thermal conductivity κ_m. The shell with radius r_2 has an anisotropic linear thermal conductivity $\bar{\kappa}_s = \text{diag}(\kappa_{rr}, \kappa_{\theta\theta})$ in cylindrical coordinates (r, θ). The core with radius r_1 owns a temperature-dependent (i.e., nonlinear) thermal conductivity $\kappa_c(T)$ given by $\kappa_c(T) = \kappa_c^{(0)} + \chi_c T^{\alpha}$, where $\kappa_c^{(0)}$ is the temperature-independent (or linear) part, χ_c and T, respectively, represent nonlinear coefficient and temperature, and α can be any real number. We assume that the core is weakly temperature-dependent (or nonlinear), say $\chi_c T^{\alpha} \ll \kappa_c^{(0)}$. Please derive the equivalent thermal conductivity $\kappa_{e2}(T)$ of the core-shell structure plus the matrix. $(\kappa_{\theta\theta}/\kappa_{rr} > 0)$.

2. Base on the above Question 1, please expand $\kappa_{e2}(T)$ in Taylor series and keep the terms up to T^{α}, namely, $\kappa_{e2}(T) = \kappa_{e2}^{(0)} + \chi_e T^{\alpha}$. Please give the concrete expression of χ_e.

Solutions

1. **Solution**: Based on Eq. (12.2), the effective thermal conductivity of the core-shell structure $\kappa_{e1}(T)$ reads

$$\kappa_{e1}(T) = u\kappa_{rr} \frac{\kappa_c(T) + u\kappa_{rr} + [\kappa_c(T) - u\kappa_{rr}]\, p_1^u}{\kappa_c(T) + u\kappa_{rr} - [\kappa_c(T) - u\kappa_{rr}]\, p_1^u},$$

and that of the core-shell structure plus the matrix $\kappa_{e2}(T)$ turns to be

$$\kappa_{e2}(T) = \kappa_m \frac{\kappa_{e1}(T) + \kappa_m + [\kappa_{e1}(T) - \kappa_m]\, p_2}{\kappa_{e1}(T) + \kappa_m - [\kappa_{e1}(T) - \kappa_m]\, p_2},$$

where $p_1 = r_1^2/r_2^2$, $p_2 = \pi r_2^2/a^2$ and $u = \sqrt{\kappa_{\theta\theta}/\kappa_{rr}}$.

2. **Solution**:

$$\chi_e = \frac{16 u^2 \kappa_{rr}^2 \kappa_m^2 p_2 p_1^u \chi_c}{\left\{ u\kappa_{rr}(p_2 - 1)\left[\kappa_c^{(0)} + u\kappa_{rr} + (\kappa_c^{(0)} - u\kappa_{rr})p_1^u\right] + \kappa_m(p_2 + 1)\left[\kappa_c^{(0)} + u\kappa_{rr} - (\kappa_c^{(0)} - u\kappa_{rr})p_1^u\right] \right\}^2}.$$

References

1. Fan, C.Z., Gao, Y., Huang, J.P.: Shaped graded materials with an apparent negative thermal conductivity. Appl. Phys. Lett. **92**, 251907 (2008)
2. Chen, T.Y., Weng, C.N., Chen, J.S.: Cloak for curvilinearly anisotropic media in conduction. Appl. Phys. Lett. **93**, 114103 (2008)
3. Guenneau, S., Amra, C., Veynante, D.: Transformation thermodynamics: cloaking and concentrating heat flux. Opt. Express **20**, 8207–8218 (2012)
4. Narayana, S., Sato, Y.: Heat flux manipulation with engineered thermal materials. Phys. Rev. Lett. **108**, 214303 (2012)
5. Schittny, R., Kadic, M., Guenneau, S., Wegener, M.: Experiments on transformation thermodynamics: molding the flow of heat. Phys. Rev. Lett. **110**, 195901 (2013)
6. Xu, H.Y., Shi, X.H., Gao, F., Sun, H.D., Zhang, B.L.: Ultrathin three-dimensional thermal cloak. Phys. Rev. Lett. **112**, 054301 (2014)
7. Han, T.C., Bai, X., Gao, D.L., Thong, J.T.L., Li, B.W., Qiu, C.-W.: Experimental demonstration of a bilayer thermal cloak. Phys. Rev. Lett. **112**, 054302 (2014)
8. Ma, Y.G., Liu, Y.C., Raza, M., Wang, Y.D., He, S.L.: Experimental demonstration of a multiphysics cloak: manipulating heat flux and electric current simultaneously. Phys. Rev. Lett. **113**, 205501 (2014)
9. Guenneau, S., Amra, C.: Anisotropic conductivity rotates heat fluxes in transient regimes. Opt. Express **21**, 6578–6583 (2013)
10. Xu, L.J., Yang, S., Huang, J.P.: Thermal theory for heterogeneously architected structure: fundamentals and application. Phys. Rev. E **98**, 052128 (2018)
11. He, X., Wu, L.Z.: Thermal transparency with the concept of neutral inclusion. Phys. Rev. E **88**, 033201 (2013)
12. Zeng, L.W., Song, R.X.: Experimental observation of heat transparency. Appl. Phys. Lett. **104**, 201905 (2014)
13. Yang, T.Z., Bai, X., Gao, D.L., Wu, L.Z., Li, B.W., Thong, J.T.L., Qiu, C.W.: Invisible sensors: simultaneous sensing and camouflaging in multiphysical fields. Adv. Mater. **27**, 7752–7758 (2015)
14. Yang, S., Xu, L.J., Wang, R.Z., Huang, J.P.: Full control of heat transfer in single-particle structural materials. Appl. Phys. Lett. **111**, 121908 (2017)
15. Wang, R.Z., Xu, L.J., Ji, Q., Huang, J.P.: A thermal theory for unifying and designing transparency, concentrating and cloaking. J. Appl. Phys. **123**, 115117 (2018)
16. Xu, L.J., Yang, S., Huang, J.P.: Thermal transparency induced by periodic interparticle interaction. Phys. Rev. Appl. **11**, 034056 (2019)
17. Han, T.C., Bai, X., Thong, J.T.L., Li, B.W., Qiu, C.-W.: Full control and manipulation of heat signatures: cloaking, camouflage and thermal metamaterials. Adv. Mat. **26**, 1731–1734 (2014)
18. Yang, T.Z., Su, Y.S., Xu, W.K., Yang, X.D.: Transient thermal camouflage and heat signature control. Appl. Phys. Lett. **109**, 121905 (2016)
19. Li, Y., Bai, X., Yang, T.Z., Luo, H., Qiu, C.W.: Structured thermal surface for radiative camouflage. Nat. Commun. **9**, 273 (2018)
20. Hu, R., Zhou, S.L., Li, Y., Lei, D.Y., Luo, X.B., Qiu, C.W.: Illusion thermotics. Adv. Mater. **30**, 1707237 (2018)
21. Zhou, S.L., Hu, R., Luo, X.B.: Thermal illusion with twinborn-like heat signatures. Int. J. Heat Mass Transfer **127**, 607 (2018)
22. Xu, L.J., Wang, R.Z., Huang, J.P.: Camouflage thermotics: a cavity without disturbing heat signatures outside. J. Appl. Phys. **123**, 245111 (2018)
23. Xu, L.J., Huang, J.P.: A transformation theory for camouflaging arbitrary heat sources. Phys. Lett. A **382**, 3313 (2018)
24. Vemuri, K.P., Bandaru, P.R.: Anomalous refraction of heat flux in thermal metamaterials. Appl. Phys. Lett. **104**, 083901 (2014)
25. Yang, T.Z., Vemuri, K.P., Bandaru, P.R.: Experimental evidence for the bending of heat flux in a thermal metamaterial. Appl. Phys. Lett. **105**, 083908 (2014)

26. Vemuri, K.P., Canbazoglu, F.M., Bandaru, P.R.: Guiding conductive heat flux through thermal metamaterials. Appl. Phys. Lett. **105**, 193904 (2014)
27. Kapadia, R.S., Bandaru, P.R.: Heat flux concentration through polymeric thermal lenses. Appl. Phys. Lett. **105**, 233903 (2014)
28. Li, Y., Shen, X.Y., Wu, Z.H., Huang, J.Y., Chen, Y.X., Ni, Y.S., Huang, J.P.: Temperature-dependent transformation thermotics: from switchable thermal cloaks to macroscopic thermal diodes. Phys. Rev. Lett. **115**, 195503 (2015)
29. Li, Y., Shen, X.Y., Huang, J.P., Ni, Y.S.: Temperature-dependent transformation thermotics for unsteady states: switchable concentrator for transient heat flow. Phys. Lett. A **380**, 1641 (2016)
30. Shen, X.Y., Li, Y., Jiang, C.R., Ni, Y.S., Huang, J.P.: Thermal cloak-concentrator. Appl. Phys. Lett. **109**, 031907 (2016)
31. Shen, X.Y., Li, Y., Jiang, C.R., Huang, J.P.: Temperature trapping: energy-free maintenance of constant temperatures as ambient temperature gradients change. Phys. Rev. Lett. **117**, 055501 (2016)
32. Dai, G.L., Shang, J., Wang, R.Z., Huang, J.P.: Nonlinear thermotics: nonlinearity enhancement and harmonic generation in thermal metasurfaces. Eur. Phys. J. B **91**, 59 (2018)
33. Nicorovici, N.A., McPhedran, R.C.: Optical and dielectric properties of partially resonant composites. Phys. Rev. B **49**, 8479 (1994)
34. Levy, O.: Nonlinear properties of partially resonant composites. J. Appl. Phys. **77**, 1696 (1995)
35. Huang, J.P., Yu, K.W.: Enhanced nonlinear optical responses of materials: composite effects. Phys. Rep. **431**, 87–172 (2006)
36. Liu, D.H., Xu, C., Hui, P.M.: Effects of a coating of spherically anisotropic material in core-shell particles. Appl. Phys. Lett. **92**, 181901 (2008)
37. Zhang, W., Ji, M., Liu, D.H.: Linear and nonlinear properties for a dilute suspension of coated ellipsoids. Phys. Lett. A **373**, 2729 (2009)
38. Zhang, W., Liu, D.H.: Second harmonic generation in composites of ellipsoidal particles with core-shell structure. Solid State Commun. **149**, 146 (2009)
39. Zeller, R.C., Pohl, R.O.: Thermal conductivity and specific heat of noncrystalline solids. Phys. Rev. B **4**, 2029–2041 (1971)
40. Glassbrenner, C.J., Slack, G.A.: Thermal conductivity of silicon and germanium from 3 K to the melting point. Phys. Rev. **134**, A1058–A1069 (1964)
41. Wegener, M.: Metamaterials beyond optics. Science **342**, 939–940 (2013)
42. Gao, Y., Huang, J.P.: Unconventional thermal cloak hiding an object outside the cloak. EPL (Europhys. Lett.) **104**, 44001 (2013)
43. Shen, X.Y., Huang, J.P.: Thermally hiding an object inside a cloak with feeling. Int. J. Heat Mass Transfer **78**, 1 (2014)
44. Xu, L.J., Yang, S., Huang, J.P.: Designing the effective thermal conductivity of materials of core-shell structure: theory and simulation. Phys. Rev. E **99**, 022107 (2019)
45. Nguyen, D.M., Xu, H.Y., Zhang, Y.M., Zhang, B.L.: Active thermal cloak. Appl. Phys. Lett. **107**, 121901 (2015)
46. Yang, S., Xu, L.J., Huang, J.P.: Metathermotics: nonlinear thermal responses of core-shell metamaterials. Phys. Rev. E **99**, 042144 (2019)

Chapter 13
Theory for Isotropic Core and Anisotropic Shell: Thermal Golden Touch

Abstract This chapter introduces the phenomenon of golden touch from myth to thermotics. We define golden touch as extending the core property to shell with extremely small core fraction. We obtain the requirement of golden touch by making the effective thermal conductivity of the core-shell structure equal to the thermal conductivity of the core. We summarize three types (A, B, and C) of golden touch in two dimensions, and only two types (A and B) of golden touch in three dimensions. We theoretically analyze the distinct properties of different types of golden touch by delicately designing the anisotropic thermal conductivity of the shell. Golden touch is also validated by finite-element simulations, which echo with the theoretical analyses. Golden touch has potential applications in thermal camouflage, thermal management, etc. This chapter not only lays the foundation for golden touch in thermotics, but also provides guidance for exploring golden touch in other diffusive fields like electrostatic and magnetostatic fields.

Keywords Thermal golden touch · Core-shell structure · Anisotropic thermal conductivity

13.1 Opening Remarks

Golden touch is a long standing dream of human beings which only exists in myth. To uncover the secret of golden touch, we should firstly define what golden touch is. We refer to the core-shell structure as our research object; see the middle structure in Fig. 13.1. For simplicity of understanding, we may imagine the shell as "stone", and the core as "gold". And then golden touch can be defined as extending the core property to shell with zero core fraction, i.e., an imaginary core. Such definition is what the golden touch in myth describes.

In spite of the difficulty, we do not give up the exploration of golden touch. Although we cannot extend the core property to shell with zero core fraction, we may resort to some loosened requirements which are physical. And hence, here we define golden touch as extending the core property to shell with extremely small core fraction. We only replace the requirement of zero core fraction with extremely small

© Springer Nature Singapore Pte Ltd. 2020
J.-P. Huang, *Theoretical Thermotics*,
https://doi.org/10.1007/978-981-15-2301-4_13

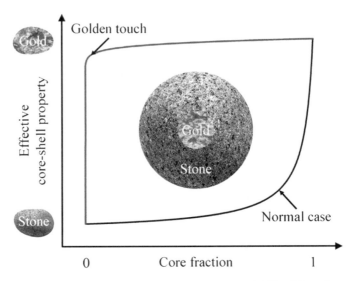

Fig. 13.1 Schematic diagram of golden touch (vs. normal case). "Gold" and "stone" are metaphors for simplicity of understanding. Adapted from Ref. [1]

core fraction. The redefinition of golden touch does not affect the inconceivable phenomenon, and makes it possible for realization. Concretely speaking, normal case only presents a slow increment with core fraction; see the black (lower right) line in Fig. 13.1. In other words, if the effective core-shell property is expected to exhibit the core property, core fraction should be 1, which echoes with the common sense of effective medium theories [2–5]. By contrast, golden touch presents a steep increment with core fraction; see the red (upper left) line in Fig. 13.1. In other words, "stone" can become "gold" with extremely small "gold" fraction. This is what we expect to obtain.

In this chapter, we focus on the thermal property of the core-shell structure, i.e., effective thermal conductivity. In fact, researches on artificial structures have realized many unique phenomena, such as thermal cloak [6–11], thermal concentrator [7, 12–15], thermal camouflage [16–22], etc. Differently, we delicately design the anisotropic shell to realize golden touch, especially when the thermal conductivity of the shell is abnormal ($\kappa_{\theta\theta}/\kappa_{rr} < 0$ for two dimensions and $\kappa_{\theta\theta}/\kappa_{rr} < -1/8$ for three dimensions). The potential application of golden touch is to dramatically reduce the use of special materials (only with extremely small core fraction). Moreover, golden touch may also provide guidance for thermal camouflage, such as size misleading.

13.2 Theory of Golden Touch

We firstly discuss the golden touch in two-dimensional core-shell structure; see Fig. 13.2a. We set the core with radius r_c and scalar thermal conductivity κ_c, and the shell with radius r_s and tensorial thermal conductivity $\kappa_s = \mathrm{diag}\,(\kappa_{rr},\ \kappa_{\theta\theta})$. We can derive the effective thermal conductivity of the core-shell structure κ_e as

$$\kappa_e\left(\kappa_{\theta\theta}/\kappa_{rr} > 0\right) = m\kappa_{rr}\,\frac{\kappa_c + m\kappa_{rr} + (\kappa_c - m\kappa_{rr})\left(\sqrt{p}\right)^{2m}}{\kappa_c + m\kappa_{rr} - (\kappa_c - m\kappa_{rr})\left(\sqrt{p}\right)^{2m}}, \tag{13.1}$$

$$\kappa_e\left(\kappa_{\theta\theta}/\kappa_{rr} < 0\right) = n\kappa_{rr}\,\frac{\kappa_c + n\kappa_{rr}\tan\left(n\ln\sqrt{p}\right)}{n\kappa_{rr} - \kappa_c\tan\left(n\ln\sqrt{p}\right)}, \tag{13.2}$$

where $m = \sqrt{\kappa_{\theta\theta}/\kappa_{rr}}$, $n = \sqrt{-\kappa_{\theta\theta}/\kappa_{rr}}$, and $p = (r_c/r_s)^2$ is the core fraction. Detailed derivation can be found in the Supplementary Proof.

We calculate the limit of Eqs. (13.1, 13.2) to discuss the property when $\kappa_{\theta\theta}/\kappa_{rr} = 0$, and find that they tend to the same value

$$\kappa_e\left(\kappa_{\theta\theta}/\kappa_{rr} = 0\right) = \kappa_{rr}\,\frac{\kappa_c - \kappa_{\theta\theta}\ln\sqrt{p}}{\kappa_{rr} - \kappa_c\ln\sqrt{p}}. \tag{13.3}$$

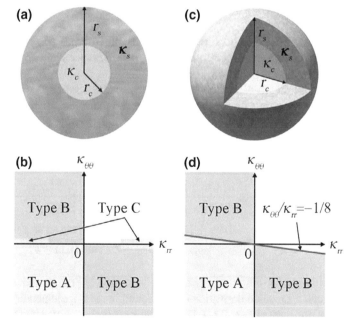

Fig. 13.2 Golden touch in **a, b** two dimensions and **c, d** three dimensions. **a, c** and **b, d** respectively present the structures and classifications of golden touch. Adapted from Ref. [1]

To be mentioned, κ_e is still dependent on $\kappa_{\theta\theta}$ despite of $\kappa_{\theta\theta}/\kappa_{rr} = 0$, because there is a condition of $\kappa_{rr} \gg \kappa_{\theta\theta} (\neq 0)$ to satisfy $\kappa_{\theta\theta}/\kappa_{rr} = 0$. Moreover, $\kappa_{\theta\theta}/\kappa_{rr} = 0$ can be regarded as the demarcation point according to Eqs. (13.1–13.3).

As the definition suggests, golden touch should firstly ensure that the core property can be extended to shell, which can be mathematically expressed as

$$\kappa_e = \kappa_c. \tag{13.4}$$

Noticing $\kappa_e (p = 0) \neq \kappa_c$, otherwise the shell is just the same as the core, which is trivial.

And secondly, the golden touch should ensure the extremely small core fraction, which can be mathematically expressed as

$$p = 0^+. \tag{13.5}$$

Eqs. (13.4, 13.5) can be regarded as the mathematical definition of golden touch.

After delicately investigating Eqs. (13.1–13.3), we find three types (A, B, and C) of golden touch in two dimensions which satisfy the requirements of Eqs. (13.4, 13.5). They are

$$\text{Type A} \rightarrow \kappa_{\theta\theta}/\kappa_{rr} > 0 : \kappa_c + m\kappa_{rr} = 0, \tag{13.6}$$

$$\text{Type B} \rightarrow \kappa_{\theta\theta}/\kappa_{rr} < 0 : \sqrt{-\kappa_{\theta\theta}/\kappa_{rr}} \ln \sqrt{p} = -Z^+\pi, \tag{13.7}$$

$$\text{Type C} \rightarrow \kappa_{\theta\theta}/\kappa_{rr} \approx 0 : \kappa_{\theta\theta} \approx 0 \text{ with } \kappa_{rr} \gg \kappa_1, \tag{13.8}$$

where $Z^+ (= 1, 2, 3, \ldots)$ is positive integers. The three types (A, B, and C) of golden touch are clearly presented in Fig. 13.2b, which respectively correspond to $>$, $<$, and $\approx$ demarcation point.

We then discuss the golden touch in three-dimensional core-shell structure; see Fig. 13.2c. We set the core with radius r_c and scalar thermal conductivity κ_c, and the shell with radius r_s and tensorial thermal conductivity $\kappa_s = \text{diag}\left(\kappa_{rr}, \kappa_{\theta\theta}, \kappa_{\varphi\varphi}\right)$ with $\kappa_{\theta\theta} = \kappa_{\varphi\varphi}$ for brevity. We can derive the effective thermal conductivity of the core-shell structure κ_e as

$$\kappa_e \left(\frac{\kappa_{\theta\theta}}{\kappa_{rr}} > -\frac{1}{8}\right) = \kappa_{rr} \frac{u_1 \left(\kappa_c - u_2\kappa_{rr}\right) - u_2 \left(\kappa_c - u_1\kappa_{rr}\right) \left(\sqrt[3]{p}\right)^{u_1-u_2}}{\left(\kappa_c - u_2\kappa_{rr}\right) - \left(\kappa_c - u_1\kappa_{rr}\right) \left(\sqrt[3]{p}\right)^{u_1-u_2}}, \tag{13.9}$$

$$\kappa_e \left(\frac{\kappa_{\theta\theta}}{\kappa_{rr}} < -\frac{1}{8}\right) = \kappa_{rr} \frac{4v\kappa_c + \left[2\kappa_c + \left(1 + 4v^2\right) \kappa_{rr}\right] \tan \left(v \ln \sqrt[3]{p}\right)}{4v\kappa_{rr} - 2 \left(2\kappa_c + \kappa_{rr}\right) \tan \left(v \ln \sqrt[3]{p}\right)}, \tag{13.10}$$

where $u_{1,2} = \left(-1 \pm \sqrt{1 + 8\kappa_{\theta\theta}/\kappa_{rr}}\right)/2, v = \sqrt{-1 - -8\kappa_{\theta\theta}/\kappa_{rr}}/2$, and $p = (r_c/r_s)^3$ is the core fraction. Detailed derivation can be found in the Supplementary Proof.

We calculate the limit of Eqs. (13.9, 13.10) to discuss the property when $\kappa_{\theta\theta}/\kappa_{rr} = -1/8$, and find that they tend to the same value

$$\kappa_e \left(\kappa_{\theta\theta}/\kappa_{rr} = -1/8\right) = \kappa_{rr} \frac{4\kappa_c + (2\kappa_c + \kappa_{rr}) \ln \sqrt[3]{p}}{4\kappa_{rr} - 2(2\kappa_c + \kappa_{rr}) \ln \sqrt[3]{p}}. \tag{13.11}$$

Here $\kappa_{\theta\theta}/\kappa_{rr} = -1/8$ can be regarded as the demarcation point according to Eqs. (13.9–13.11).

We also calculate the effective thermal conductivity when $\kappa_{\theta\theta}/\kappa_{rr} = 0$ as a special case

$$\kappa_e \left(\kappa_{\theta\theta}/\kappa_{rr} = 0\right) = \kappa_{rr} \frac{\kappa_c \sqrt[3]{p}}{\kappa_{rr} + \kappa_c \left(1 - \sqrt[3]{p}\right)}. \tag{13.12}$$

Here κ_e is independent of $\kappa_{\theta\theta}$ which is different from the two-dimensional result of Eq. (13.3).

According to the mathematical definition of golden touch Eqs. (13.4, 13.5), we delicately investigate Eqs. (13.9–13.12), but only find two types (A and B) of golden touch. They are

$$\text{Type A} \rightarrow \frac{\kappa_{\theta\theta}}{\kappa_{rr}} > -\frac{1}{8} : \kappa_c - u_2\kappa_{rr} = 0, \tag{13.13}$$

$$\text{Type B} \rightarrow \frac{\kappa_{\theta\theta}}{\kappa_{rr}} < -\frac{1}{8} : \left(\frac{\sqrt{-1 - -8\kappa_{\theta\theta}/\kappa_{rr}}}{2}\right) \ln \sqrt[3]{p} = -Z^+\pi, \tag{13.14}$$

where $Z^+ (= 1, 2, 3, \ldots)$ is positive integers. The two types (A and B) of golden touch are clearly presented in Fig. 13.2d, which respectively correspond to $>$ and $<$ demarcation point. Different from two-dimensional system, there is no type C golden touch in three-dimensional system, even though we carefully calculate the effective thermal conductivity of the core-shell structure when $\kappa_{\theta\theta}/\kappa_{rr} = -1/8$ (demarcation point in three dimensions; Eq. 13.11), or $\kappa_{\theta\theta}/\kappa_{rr} = 0$ (Eq. 13.12).

13.3 Theoretical Analyses of Golden Touch

We further analyze the distinct properties of different types of golden touch. For clarity, we discuss the dimensionless thermal conductivity κ_e/κ_c. When $\kappa_e/\kappa_c = 1$, the core property is extended to shell.

The two-dimensional results of type A, type B, and type C golden touch are respectively demonstrated in Fig. 13.3a–i. We will give detailed discussions of the three types of golden touch in the following.

Type A: When the requirement of Eq. (13.6) is strictly satisfied, there is a discontinuous change from -1 to 1 at $p = 0$; see Fig. 13.3a. In other words, the core property can be extended to shell with arbitrarily small core fraction once $p \neq 0$. We increase the thermal conductivity of the shell; see Fig. 13.3b, and the variation curves become similar to a hyperbolic function. When the increment is small enough, golden touch still works; see the green (lightest) line in Fig. 13.3b. However, when

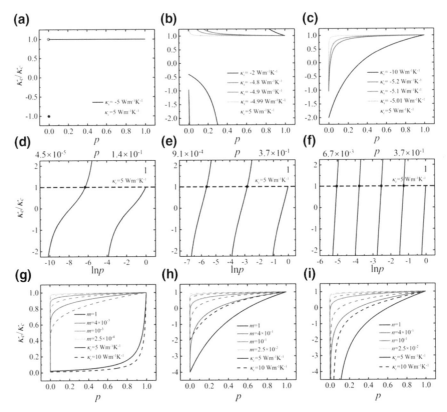

Fig. 13.3 Theoretical analyses of two-dimensional golden touch for **a–c** type A, **d–f** type B, and **g–i** type C. Concrete parameters: **a–c** $\kappa_c = 5\,\mathrm{Wm^{-1}K^{-1}}$, and the thermal conductivity of the shell is a scalar denoted as κ_s; **d–f** $\kappa_c = 5\,\mathrm{Wm^{-1}K^{-1}}$, $\kappa_{rr} = 5\,\mathrm{Wm^{-1}K^{-1}}$, **d** $n = 1$, **e** $n = \sqrt{5}$, and **f** $n = 5$; **g–i** $\kappa_c = 5\,\mathrm{Wm^{-1}K^{-1}}$ for solid lines, $\kappa_c = 10\,\mathrm{Wm^{-1}K^{-1}}$ for dashed lines, **g** $m\kappa_{rr} = 0.1\,\mathrm{Wm^{-1}K^{-1}}$, **h** $m\kappa_{rr} = -20\,\mathrm{Wm^{-1}K^{-1}}$, and **i** $n\kappa_{rr} = 20\,\mathrm{Wm^{-1}K^{-1}}$. It should be noted that there are three lines in the left bottom of (**b**), which are very close to each other. Adapted from Ref. [1]

the increment is big, golden touch turns into the normal case; see the black (darkest) line in Fig. 13.3b. We decrease the thermal conductivity of the shell; see Fig. 13.3c, and the variation curves become monotonically increasing without a discontinuous change. When the decrement is small enough, golden touch still works; see the green (lightest) line in Fig. 13.3c. However, when the decrement is big, golden touch turns into the normal case; see the black (darkest) line in Fig. 13.3c. Therefore, type A golden touch can work perfectly with arbitrarily small core fraction, but requires a special relation of thermal conductivities between the core and shell.

Type B: In fact, as long as $\kappa_{\theta\theta}/\kappa_{rr} < 0$, the phenomena of golden touch will exist, for the curve value ranges from $-\infty$ to $+\infty$ and presents quasi-periodicity; see Fig. 13.3d–f. The quasi-periodicity is determined by the shell anisotropy: from left to right of Fig. 13.3d–f, the smaller $\kappa_{\theta\theta}/\kappa_{rr}$ is (or the bigger n is), the denser quasi-periodicity is. Therefore, type B golden touch can work perfectly without requirement of thermal conductivities between the core and shell, but with certain core fraction determined by Eq. (13.7); see the dots in Fig. 13.3d–f. To be mentioned, we use ln p (ranging from $-\infty$ to 0) as the abscissa to show the infinite numbers of quasi-periodicity, and hence the core fraction can also be set as arbitrarily small.

Type C: The parameters of Eq. (13.8) are distributed in four quadrants; see Fig. 13.2b. When $\kappa_{\theta\theta}/\kappa_{rr}$ is in the third quadrant, type C golden touch possesses all the properties of type A golden touch. When $\kappa_{\theta\theta}/\kappa_{rr}$ is in the second (or forth) quadrant, type C golden touch possesses all the properties of type B golden touch. Even so, we still regard type C golden touch as a separate classification, for it possesses different properties from type A and type B golden touch. Concretely speaking, type A and type B golden touch require certain thermal conductivity (Eq. 13.6) or core fraction (Eq. 13.7), but these requirements disappear in type C golden touch. In other words, any thermal conductivity can be extended with any core fraction. The costs are (I) the requirement of Eq. (13.8), and (II) the core fraction can only be extremely small rather than arbitrarily small, which is dependent on the shell. We take the parameters in the first quadrant as an example; see Fig. 13.3g. Solid and dashed lines respectively correspond to the different thermal conductivities of cores. With the decrement of $m\left(=\sqrt{\kappa_{\theta\theta}/\kappa_{rr}}\rightarrow 0\right)$, the core property can be extended to shell regardless of the core conductivities; see the green (lightest) solid and dashed lines in Fig. 13.3g. We also investigate parameters in the third and forth quadrants, and the results are respectively shown in Fig. 13.3h, i which are similar to those in Fig. 13.3g. To be mentioned, the reason why the variation curves in Fig. 13.3i do not present the quasi-periodicity with $\kappa_{\theta\theta}/\kappa_{rr} < 0$ is that the quasi-periodicity becomes sparse with small n (as discussed in type B golden touch), and thus exists in extremely small core fraction, which cannot be shown under the abscissa of p.

The three-dimensional results of type A and type B golden touch are respectively demonstrated in Fig. 13.4a–f. Except for that there is no type C golden touch in three dimensions, type A and type B golden touch in three dimensions are similar to those in two dimensions.

Type A: When the requirement of Eq. (13.13) is strictly satisfied, there is a discontinuous change from -0.5 to 1 at $p = 0$; see Fig. 13.4a. We increase the thermal conductivity of the shell; see Fig. 13.4b, and the variation curves become similar to a hyperbolic function. When the increment is small enough, golden touch still works; see the green (lightest) line in Fig. 13.4b. However, when the increment is big, golden touch turns into the normal case; see the black (darkest) line in Fig. 13.4b. We decrease the thermal conductivity of the shell; see Fig. 13.4c, and the variation curves become monotonically increasing without a discontinuous change. When the decrement is small enough, golden touch still works; see the green (lightest) line in Fig. 13.4c. However, when the decrement is big, golden touch turns into the normal case; see the black (darkest) line in Fig. 13.4c. Therefore, type A golden touch can work perfectly with arbitrarily small core fraction, but requires a special relation of thermal conductivities between the core and shell.

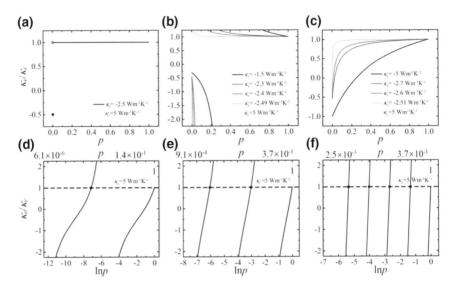

Fig. 13.4 Theoretical analyses of three-dimensional golden touch for **a–c** type A and **d–f** type B. Concrete parameters: **a–c** $\kappa_c = 5\,\mathrm{Wm^{-1}K^{-1}}$, and the thermal conductivity of the shell is a scalar denoted as κ_s; **d–f** $\kappa_c = 5\,\mathrm{Wm^{-1}K^{-1}}$, $\kappa_{rr} = 5\,\mathrm{Wm^{-1}K^{-1}}$, **d** $v = \sqrt{7}/2$, **e** $v = \sqrt{39}/2$, and **f** $n = \sqrt{199}/2$. Adapted from Ref. [1]

Type B: As long as $\kappa_{\theta\theta}/\kappa_{rr} < -1/8$, the phenomena of golden touch will exist, for the curve value ranges from $-\infty$ to $+\infty$ and present quasi-periodicity; see Fig. 13.4d–f. The quasi-periodicity is determined by the shell anisotropy: from left to right of Fig. 13.4d–f, the smaller $\kappa_{\theta\theta}/\kappa_{rr}$ is (or the bigger v is), the denser quasi-periodicity is. Therefore, type B golden touch can work perfectly without requirement of thermal conductivities between the core and shell, but with certain core fraction determined by Eq. (13.14); see the dots in Fig. 13.4d–f. We also use $\ln p$ (ranging from $-\infty$ to 0) as the abscissa to show the infinite numbers of quasi-periodicity, and hence the core fraction can also be set as arbitrarily small.

13.4 Finite-Element Simulations of Golden Touch

We have theoretically analyzed the distinct properties of different types of golden touch in both two and three dimensions. Now we are in the position to demonstrate finite-element simulations for intuitive understanding of golden touch. We put the core-shell structure into a matrix (κ_m) with the same thermal conductivity of the core ($\kappa_m = \kappa_c$). If golden touch does extend the core property to shell, the external thermal field will keep unchanged, namely uniform temperature gradient. To be mentioned, although the core fraction of type A and type B golden touch can be arbitrarily small, we have to set the core fraction as a reasonable finite small value

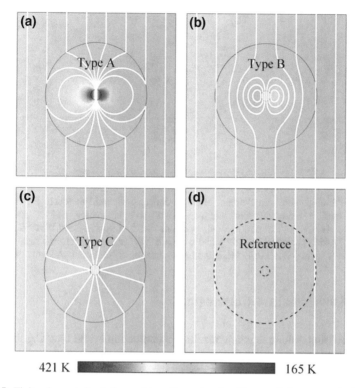

421 K ████████████████████████ 165 K

Fig. 13.5 Finite-element simulations of two-dimensional golden touch for **a** type A, **b** type B, **c** type C, and **d** reference. Concrete parameters: **a–d** simulation box is 20×20 cm, $r_s = 6.4$ cm, $r_c = 0.64$ cm, $\kappa_c = \kappa_m = 5$ Wm^{-1}K^{-1}, **a** $\kappa_s = -5.001$ Wm^{-1}K^{-1}, **b** $\kappa_s = \text{diag}(5, -9.308)$ Wm^{-1}K^{-1}, **c** $\kappa_s = \text{diag}(400, 2.5 \times 10^{-6})$ Wm^{-1}K^{-1}, and **d** $\kappa_s = 5$ Wm^{-1}K^{-1}. Dashed lines in **d** are used to show the imaginary location of the core-shell structure for simplicity of comparison. The left and right boundaries are respectively set at 313 and 273 K, and other boundaries are insulated. White lines represent isotherms. Adapted from Ref. [1]

to perform finite-element simulations based on the commercial software COMSOL Multiphysics (http://www.comsol.com/).

The results of two-dimensional golden touch are presented in Fig. 13.5. We set the core fraction as 0.01. Type A, type B, and type C golden touch are respectively designed according to Eqs. (13.6–13.8). For type C golden touch is not an exact result (extremely small rather than arbitrarily small), the parameters applied for finite-element simulation echo with the green (lightest) solid line in Fig. 13.3g. The same uniform temperature gradient between matrix in Figs. 13.5a–d validates the theoretically-predicted golden touch.

The results of three-dimensional golden touch are presented in Fig. 13.6. We set the core fraction as 0.008. Type A and type B golden touch are respectively designed according to Eqs. (13.13, 13.14). It is found that the external temperature distribution in Fig. 13.6a–c is totally the same, which validates the golden touch again.

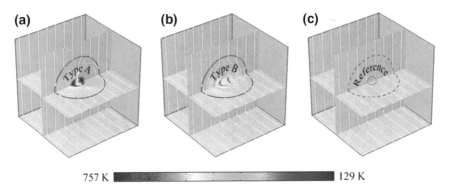

Fig. 13.6 Finite-element simulations of three-dimensional golden touch for **a** type A, **b** type B, and **c** reference. Concrete parameters: **a–c** simulation box is $20 \times 20 \times 20\,\mathrm{cm}$, $r_s = 6.4\,\mathrm{cm}$, $r_c = 1.28\,\mathrm{cm}$, $\kappa_c = \kappa_m = 5\,\mathrm{Wm^{-1}K^{-1}}$, **a** $\kappa_s = -2.501\,\mathrm{Wm^{-1}K^{-1}}$, **b** $\kappa_s =$ diag$(-10, 77.455, 77.455)\,\mathrm{Wm^{-1}K^{-1}}$, and **c** $\kappa_s = 5\,\mathrm{Wm^{-1}K^{-1}}$. The left and right boundaries are respectively set at 463 K and 423 K, and other boundaries are insulated. Adapted from Ref. [1]

13.5 Discussion and Conclusion

When discussing golden touch, a puzzling phenomenon is that type C golden touch cannot be extended from two dimensions to three dimensions. In other words, type C golden touch is a unique phenomenon which only exists in two dimensions. In fact, low-dimensional heat transfer at microscopic scale has been found many unique properties, such as the nonconvergence effect and size effect of the thermal conductivity [23]. However, the uniqueness of low-dimensional heat transfer has never been discovered at macroscopic scale, such as thermal metamaterials including but not limited to thermal cloak [6–11], thermal concentrator [7, 12–15], and thermal camouflage [16–22]. Therefore, type C golden touch might open a gate to explore unique properties in low-dimensional heat transfer at macroscopic scale.

Moreover, type C golden touch in the first quadrant also seems to be distinct, for it is the only golden touch which requires no apparent negative thermal conductivity; see Fig. 13.2b–d. Although apparent negative thermal conductivity does not exist in nature, it can be realized by active materials containing heat sources [24–26]. We take two-dimensional type A golden touch (Fig. 13.5a) as an example. We set the thermal conductivity of the shell with a positive value which is different from that of the core. To realize the same effect of golden touch, we add continuous sources (Fig. 13.7a) and discontinuous sources (Fig. 13.7b) on the boundaries of the shell. The same temperature profile between Figs. 13.5a and 13.7a, b validates that the scheme of adding sources works indeed. For experimental realization, Ref. [27] demonstrates a device to realize the discontinuous sources, which makes apparent negative thermal conductivities feasible for experiments.

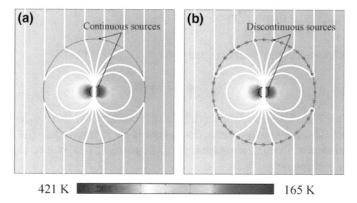

Fig. 13.7 Realization of apparent negative thermal conductivity by adding **a** continuous sources and **b** discontinuous sources. The parameters are the same as those in Fig. 13.5a, except for the shell which is set as $\kappa_s = 1\,\mathrm{Wm^{-1}K^{-1}}$. **a** Continuous sources are applied on the inner and outer boundaries of the designed shell which obey the continuous temperature distribution $T = -81.92x/r_{c,s}^2 + 293$, where $r_c = 0.64$ cm ($r_s = 6.4$ cm) is the inner (or outer) radius of the designed shell. x represents abscissa whose origin locates in the center of the simulation box. For the inner boundary, x ranges from $-r_c$ to r_c which makes the temperature T range from 421 to 165 K; for the outer boundary, x ranges from $-r_s$ to r_s which makes temperature T range from 306 to 280 K. **b** Twelve discontinuous sources (with radius 0.05 cm) and thirty-six discontinuous sources (with radius 0.15 cm) are respectively applied on the inner and outer boundaries. The discontinuous temperatures are calculated from the continuous temperature distribution in **a** according to the source abscissas. Adapted from Ref. [1]

We also first derive the effective thermal conductivity under the demarcation point, i.e., $\kappa_{\theta\theta}/\kappa_{rr} < 0$ for two dimensions Eq. (13.2), and $\kappa_{\theta\theta}/\kappa_{rr} < -1/8$ for three dimensions Eq. (13.10). This helps to reveal the quasi-periodic variation with core fraction; see Figs. 13.3d–f and 13.4d–f, which is dramatically different from the well-known effective medium theories like the Maxwell-Garnett formula [28] and the Bruggeman formula [29]. This may further provide guidance for exploring non-linear effects [30] beyond the framework of Maxwell-Garnett formula or Bruggeman formula. Moreover, one reliable approach to realize these special thermal conductivities is to design multilayer structures with effective medium theory. In this way, the complex parameters can be obtained with several homogeneous and isotropic materials which are easy to get.

In summary, golden touch proposed in this chapter can extend the core property to shell with extremely small core fraction, which has potential applications in thermal camouflage, thermal management, etc. Furthermore, golden touch can be directly extended to electrostatics and magnetostatics where permittivity and permeability play the same role as thermal conductivity in thermotics. Golden touch in magnetostatics may also offer guidance for magnetostatic camouflage [31–33].

13.6 Supplementary Proof

The dominant equation of heat conduction is

$$\nabla \cdot (-\kappa \nabla T) = 0, \tag{13.15}$$

where κ and T are respectively tensorial thermal conductivity and temperature.

We firstly discuss the two-dimensional core-shell structure, and put it into an infinite matrix with thermal conductivity κ_m. Equation (13.15) can be expanded in cylindrical coordinates as

$$\frac{\partial}{\partial r} \left(r \kappa_{rr} \frac{\partial T}{\partial r} \right) + \frac{\partial}{\partial \theta} \left(\kappa_{\theta\theta} \frac{\partial T}{r \partial \theta} \right) = 0. \tag{13.16}$$

The general solution of Eq. (13.16) is

$$T \left(\kappa_{\theta\theta}/\kappa_{rr} > 0 \right) = A_0 + B_0 \ln r + \sum_{i=1}^{\infty} [A_i \sin(i\theta) + B_i \cos(i\theta)] r^{im_1}$$

$$+ \sum_{i=1}^{\infty} [C_i \sin(i\theta) + D_i \cos(i\theta)] r^{im_2}, \tag{13.17}$$

$$T \left(\kappa_{\theta\theta}/\kappa_{rr} < 0 \right) = E_0 + F_0 \ln r + \sum_{i=1}^{\infty} [E_i \sin(i\theta) + F_i \cos(i\theta)] \sin(in \ln r)$$

$$+ \sum_{i=1}^{\infty} [G_i \sin(i\theta) + H_i \cos(i\theta)] \cos(in \ln r), \tag{13.18}$$

where $m_{1,2} = \pm\sqrt{\kappa_{\theta\theta}/\kappa_{rr}}$, and $n = \sqrt{-\kappa_{\theta\theta}/\kappa_{rr}}$. Here $\kappa_{\theta\theta}/\kappa_{rr} = 0$ is the demarcation point.

The temperature distribution of the core (T_c), shell (T_s), and matrix (T_m) can then be determined by the following boundary conditions,

$$\begin{cases} T_c < \infty, \\ T_c(r_c) = T_s(r_c), \\ T_s(r_s) = T_m(r_s), \\ (-\kappa_c \partial T_c/\partial r)_{r_c} = (-\kappa_{rr} \partial T_s/\partial r)_{r_c}, \\ (-\kappa_{rr} \partial T_s/\partial r)_{r_s} = (-\kappa_m \partial T_m/\partial r)_{r_s}, \\ \nabla T_m(r \to \infty) = \nabla T_0, \end{cases} \tag{13.19}$$

where ∇T_0 represents the external uniform thermal field gradient.

For the symmetric core-shell structure and boundary conditions, we only require to keep several terms of $i = 1$ in Eqs. (13.17, 13.18),

$$T\left(\kappa_{\theta\theta}/\kappa_{rr} > 0\right) = A_0 + B_1 r^{m_1} \cos\theta + D_1 r^{m_2} \cos\theta, \tag{13.20}$$

$$T\left(\kappa_{\theta\theta}/\kappa_{rr} < 0\right) = E_0 + F_1 \cos\theta \sin\left(n \ln r\right) + H_1 \cos\theta \cos\left(n \ln r\right). \tag{13.21}$$

Therefore, for isotropic matrix, we can obtain $T_m = A_0 + B_1 r \cos\theta + D_1 r^{-1} \cos\theta$. We set D_1 as zero to ensure the external thermal field undistorted. Then we can derive the effective thermal conductivity of the core-shell structure κ_e as Eqs. (13.1, 13.2).

We secondly discuss the three-dimensional core-shell structure, and also put it into an infinite matrix with thermal conductivity κ_m. Equation (13.15) can be expanded in spherical coordinates as

$$\frac{1}{r}\frac{\partial}{\partial r}\left(r^2 \kappa_{rr}\frac{\partial T}{\partial r}\right) + \frac{1}{\sin\theta}\frac{\partial}{\partial\theta}\left(\sin\theta\,\kappa_{\theta\theta}\frac{\partial T}{r\partial\theta}\right) = 0. \tag{13.22}$$

The general solution of Eq. (13.22) is

$$T\left(\kappa_{\theta\theta}/\kappa_{rr} \geq 0\right) = A_0 + B_0 r^{-1} + \sum_{i=1}^{\infty}\left(A_i r^{s_1} + B_i r^{s_2}\right) P_i\left(\cos\theta\right), \tag{13.23}$$

$$T\left(0 > \kappa_{\theta\theta}/\kappa_{rr} > -1/8\right) = C_0 + D_0 r^{-1} + \sum_{i=1}^{j}\left(C_i r^{s_1} + D_i r^{s_2}\right) P_i\left(\cos\theta\right)$$
$$+ \sum_{i=j+1}^{\infty} r^{-1/2}\left[E_i \sin\left(t \ln r\right) + F_i \cos\left(t \ln r\right)\right] P_i\left(\cos\theta\right), \tag{13.24}$$

$$T\left(\kappa_{\theta\theta}/\kappa_{rr} < -1/8\right) = G_0 + H_0 r^{-1}$$
$$+ \sum_{i=1}^{\infty} r^{-1/2}\left[G_i \sin\left(t \ln r\right) + H_i \cos\left(t \ln r\right)\right] P_i\left(\cos\theta\right), \tag{13.25}$$

where $s_{1,2} = \left(-1 \pm \sqrt{1 + 4i\left(i+1\right)\kappa_{\theta\theta}/\kappa_{rr}}\right)/2$, $t = \sqrt{-1 - -4i\left(i+1\right)\kappa_{\theta\theta}/\kappa_{rr}}/2$, and $j = \text{INT}\left[\left(-1 + \sqrt{1 - \kappa_{rr}/\kappa_{\theta\theta}}\right)/2\right]$, where i is the summation index in Eqs. (13.23–13.25), and $\text{INT}\left[\cdots\right]$ is the integral function with respect to .... P_i is Legendre polynomials.

We find that Eqs. (13.23, 13.24) are essentially the same with similar boundary conditions of Eq. (13.19), for we only require to keep several terms of $i = 1$,

$$T\left(\frac{\kappa_{\theta\theta}}{\kappa_{rr}} > -\frac{1}{8}\right) = A_0 + \left(A_1 r^{s_1} + B_1 r^{s_2}\right)\cos\theta, \tag{13.26}$$

$$T\left(\frac{\kappa_{\theta\theta}}{\kappa_{rr}} < -\frac{1}{8}\right) = G_0 + r^{-1/2}\left[G_1 \sin\left(t \ln r\right) + H_1 \cos\left(t \ln r\right)\right]\cos\theta. \tag{13.27}$$

Therefore, $\kappa_{\theta\theta}/\kappa_{rr} = -1/8$ is the real demarcation point. For isotropic matrix, we can obtain $T_m = A_0 + \left(A_1 r + B_1 r^{-2}\right)\cos\theta$. We set B_1 as zero to ensure the external thermal field undistorted. Then we can derive the effective thermal conductivity of the core-shell structure κ_e as Eqs. (13.9, 13.10).

References

1. Xu, L.J., Yang, S., Huang, J.P.: Designing the effective thermal conductivity of materials of core-shell structure: theory and simulation. Phys. Rev. E **99**, 022107 (2019)
2. Bergman, D.J., Stroud, D.: Physical properties of macroscopically inhomogeneous media. Solid State Phys. **46**, 147–269 (1992)
3. Huang, J.P., Yu, K.W.: Enhanced nonlinear optical responses of materials: composite effects. Phys. Rep. **431**, 87–172 (2006)
4. Yang, S., Xu, L.J., Wang, R.Z., Huang, J.P.: Full control of heat transfer in single-particle structural materials. Appl. Phys. Lett. **111**, 121908 (2017)
5. Xu, L.J., Jiang, C.R., Shang, J., Wang, R.Z., Huang, J.P.: Periodic composites: Quasi-uniform heat conduction, Janus thermal illusion, and illusion thermal diodes. Eur. Phys. J. B **90**, 221 (2017)
6. Fan, C.Z., Gao, Y., Huang, J.P.: Shaped graded materials with an apparent negative thermal conductivity. Appl. Phys. Lett. **92**, 251907 (2008)
7. Narayana, S., Sato, Y.: Heat flux manipulation with engineered thermal materials. Phys. Rev. Lett. **108**, 214303 (2012)
8. Han, T.C., Yuan, T., Li, B.W., Qiu, C.-W.: Homogeneous thermal cloak with constant conductivity and tunable heat localization. Sci. Rep. **3**, 1593 (2013)
9. Xu, H.Y., Shi, X.H., Gao, F., Sun, H.D., Zhang, B.L.: Ultrathin three-dimensional thermal cloak. Phys. Rev. Lett. **112**, 054301 (2014)
10. Han, T.C., Bai, X., Gao, D.L., Thong, J.T.L., Li, B.W., Qiu, C.-W.: Experimental demonstration of a bilayer thermal cloak. Phys. Rev. Lett. **112**, 054302 (2014)
11. Ma, Y.G., Liu, Y.C., Raza, M., Wang, Y.D., He, S.L.: Experimental demonstration of a multiphysics cloak: manipulating heat flux and electric current simultaneously. Phys. Rev. Lett. **113**, 205501 (2014)
12. Han, T.C., Zhao, J.J., Yuan, T., Lei, D.Y., Li, B.W., Qiu, C.W.: Theoretical realization of an ultra-efficient thermalenergy harvesting cell made of natural materials. Energ. Environ. Sci. **6**, 3537–3541 (2013)
13. Moccia, M., Castaldi, G., Savo, S., Sato, Y., Galdi, V.: Independent manipulation of heat and electrical current via bifunctional metamaterials. Phys. Rev. X **4**, 021025 (2014)
14. Chen, T.Y., Weng, C.N., Tsai, Y.L.: Materials with constant anisotropic conductivity as a thermal cloak or concentrator. J. Appl. Phys. **117**, 054904 (2015)
15. Shen, X.Y., Li, Y., Jiang, C.R., Ni, Y.S., Huang, J.P.: Thermal cloak-concentrator. Appl. Phys. Lett. **109**, 031907 (2016)
16. Han, T.C., Bai, X., Thong, J.T.L., Li, B.W., Qiu, C.-W.: Full control and manipulation of heat signatures: cloaking, camouflage and thermal metamaterials. Adv. Mat. **26**, 1731–1734 (2014)
17. Yang, T.Z., Bai, X., Gao, D.L., Wu, L.Z., Li, B.W., Thong, J.T.L., Qiu, C.W.: Invisible sensors: simultaneous sensing and camouflaging in multiphysical fields. Adv. Mater. **27**, 7752–7758 (2015)
18. Yang, T.Z., Su, Y.S., Xu, W.K., Yang, X.D.: Transient thermal camouflage and heat signature control. Appl. Phys. Lett. **109**, 121905 (2016)
19. Xu, L.J., Wang, R.Z., Huang, J.P.: Camouflage thermotics: a cavity without disturbing heat signatures outside. J. Appl. Phys. **123**, 245111 (2018)
20. Xu, L.J., Huang, J.P.: A transformation theory for camouflaging arbitrary heat sources. Phys. Lett. A **382**, 3313 (2018)
21. Hu, R., Zhou, S.L., Li, Y., Lei, D.Y., Luo, X.B., Qiu, C.W.: Illusion thermotics. Adv. Mater. **30**, 1707237 (2018)
22. Zhou, S.L., Hu, R., Luo, X.B.: Thermal illusion with twinborn-like heat signatures. Int. J. Heat Mass Transf. **127**, 607 (2018)
23. Lepri, S., Livi, R., Politi, A.: Thermal conduction in classical low-dimensional lattices. Phys. Rep. **377**, 1 (2003)
24. Wegener, M.: Metamaterials beyond optics. Science **342**, 939–940 (2013)

25. Gao, Y., Huang, J.P.: Unconventional thermal cloak hiding an object outside the cloak. EPL (Europhys. Lett.) **104**, 44001 (2013)
26. Shen, X.Y., Huang, J.P.: Thermally hiding an object inside a cloak with feeling. Int. J. Heat Mass Transf. **78**, 1 (2014)
27. Nguyen, D.M., Xu, H.Y., Zhang, Y.M., Zhang, B.L.: Active thermal cloak. Appl. Phys. Lett. **107**, 121901 (2015)
28. Garnett, J.C.M.: Colours in metal glasses and in metallic films. Philos. Trans. R. Soc. London Ser. A **203**, 385 (1904)
29. Bruggeman, D.A.G.: Berechnung verschiedener physikalischer Konstanten von heterogenen substanzen. I. Dielektrizitätskonstanten und Leitfähigkeiten der Mischkörper aus isotropen Substanzen (Calculation of different physical constants of heterogeneous substances. I. Dielectricity and conductivity of mixtures of isotropic substances). Annalen der Physik **24**, 636–664 (1935)
30. Dai, G.L., Shang, J., Wang, R.Z., Huang, J.P.: Nonlinear thermotics: nonlinearity enhancement and harmonic generation in thermal metasurfaces. Eur. Phys. J. B **91**, 59 (2018)
31. Gomory, F., Solovyov, M., Souc, J., Navau, C., Camps, J.P., Sanchez, A.: Experimental realization of a magnetic cloak. Science **335**, 1466–1468 (2012)
32. Batlle, R.M., Parra, A., Laut, S., Valle, N.D., Navau, C., Sanchez, A.: Magnetic illusion: transforming a magnetic object into another object by negative permeability. Phys. Rev. Appl. **9**, 034007 (2018)
33. Jiang, W., Ma, Y.G., He, S.L.: Static magnetic cloak without a superconductor. Phys. Rev. Appl. **9**, 054041 (2018)

Chapter 14
Theory for Isotropic Core and Anisotropic Shell or for Two Isotropic Shells: Thermal Chameleon

Abstract Intelligence has become one of the developing trends of thermal metamaterials in order to meet different practical requirements. By considering the temperature-dependent and specially-designed thermal conductivities, chameleon-like behaviors have been revealed to realize adaptive responses to nearby objects. However, the existing schemes are approximately valid only for a small working range of nearby thermal conductivities. This fact limits practical applications. To solve this problem, here we propose two exact schemes to realize thermal chameleon-like behaviors, say, monolayer schemes and bilayer schemes. By carefully designing the thermal conductivities of the metashells, we find that the effective thermal conductivities can exactly change with those of nearby objects. In both schemes, apparent negative thermal conductivities are required, which can be realized by adding external heat sources. Theoretical derivations are validated by finite-element simulations. We further extend the monolayer schemes to three dimensions. The proposed schemes can work as a type of multifunction materials to meet different requirements of thermal conductivities. This chapter provides intelligence to thermal conductivities, which may inspire further development of intelligent thermal metamaterials.

Keywords Thermal chameleon · Monolayer schemes · Bilayer schemes · Intelligent thermal metamaterial

14.1 Opening Remarks

Thermal metamaterials have made a considerable impact on the field of heat management due to their specially-designed structures and conductivities. Some representative examples are thermal cloaks [1–10], thermal transparency [11–14], thermal bending [15–18], thermal camouflage/illusion [19–25], thermal Janus structures [26], etc.

However, these schemes almost exhibit no intelligence, which means that they cannot adapt to the change of nearby objects. Recent works have made some advances in intelligence, and designed chameleonlike metashells in different diffusion fields [27–29]. However, these results are almost approximately valid, which rely on a small working range of nearby objects.

© Springer Nature Singapore Pte Ltd. 2020
J.-P. Huang, *Theoretical Thermotics*,
https://doi.org/10.1007/978-981-15-2301-4_14

To solve this problem, here we propose two exact schemes to realize metashells with thermal chameleonlike behaviors. Concretely speaking, the designed metashells can always imitate the thermal conductivity of nearby objects, thus yielding an undistorted temperature profile outside the metashell (as if the metashell were the same as the objects in the vicinity). Our method depends on the exact derivation of the effective thermal conductivity of core-shell structures.

14.2 Theory for Thermal Chameleonlike Metashells

14.2.1 Anisotropic Monolayer Schemes

We discuss the properties of the core-shell structure presented in Fig. 14.1a. We set the radius and the thermal conductivity of the core to be r_c and κ_c, and those of the shell to be r_s and $\kappa_s = \mathrm{diag}\,(\kappa_{rr},\ \kappa_{\theta\theta})$ with $\kappa_{\theta\theta}/\kappa_{rr} < 0$ in cylindrical coordinates $(r,\ \theta)$. The dominant equation in heat conduction is

$$\nabla \cdot (-\kappa \nabla T) = 0, \tag{14.1}$$

where κ and T are tensorial thermal conductivity and temperature, respectively.
    Equation (14.1) can be expressed in cylindrical coordinates as

$$\frac{\partial}{\partial r}\left(r\kappa_{rr}\frac{\partial T}{\partial r}\right) + \frac{\partial}{\partial \theta}\left(\kappa_{\theta\theta}\frac{\partial T}{r\partial \theta}\right) = 0. \tag{14.2}$$

The general solution of Eq. (14.2) is

$$T\,(\kappa_{\theta\theta}/\kappa_{rr} < 0) = A_0 + B_0 \ln r$$

$$+ \sum_{i=1}^{\infty}[A_i \sin\,(i\theta) + B_i \cos\,(i\theta)]\sin\,(im \ln r)$$

$$+ \sum_{i=1}^{\infty}[C_i \sin\,(i\theta) + D_i \cos\,(i\theta)]\cos\,(im \ln r)\,, \tag{14.3}$$

$$T\,(\kappa_{\theta\theta}/\kappa_{rr} > 0) = E_0 + F_0 \ln r$$

$$+ \sum_{i=1}^{\infty}[E_i \sin\,(i\theta) + F_i \cos\,(i\theta)]\,r^{in}$$

$$+ \sum_{i=1}^{\infty}[G_i \sin\,(i\theta) + H_i \cos\,(i\theta)]\,r^{-in}, \tag{14.4}$$

where $m = \sqrt{-\kappa_{\theta\theta}/\kappa_{rr}}$, and $n = \sqrt{\kappa_{\theta\theta}/\kappa_{rr}}$.

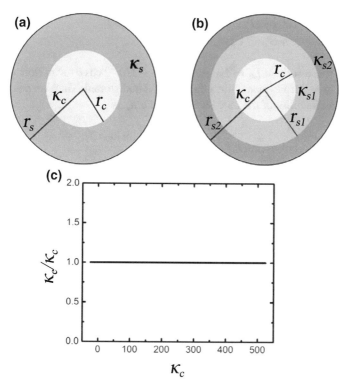

Fig. 14.1 Schematic diagrams of **a** core-shell structure or **b** core-shell-shell structure in two dimensions. **c** Shows the functional relationship between κ_e/κ_c and κ_c when the thermal chameleonlike metashells work. Adapted from Ref. [33]

The temperature distributions of the core (T_c), the shell (T_s), and the matrix (T_m) can then be determined by the following boundary conditions,

$$\begin{cases}
T_c < \infty, \\
T_c(r_c) = T_s(r_c), \\
T_s(r_s) = T_m(r_s), \\
(-\kappa_c \partial T_c/\partial r)_{r_c} = (-\kappa_{rr} \partial T_s/\partial r)_{r_c}, \\
(-\kappa_{rr} \partial T_s/\partial r)_{r_s} = (-\kappa_m \partial T_m/\partial r)_{r_s}, \\
\nabla T_m(r \to \infty) = \nabla T_0,
\end{cases} \tag{14.5}$$

where ∇T_0 represents the external uniform potential gradient.

We only require to keep several terms with $i = 1$ in Eqs. (14.3) and (14.4) because of the symmetric boundary conditions of Eq. (14.5),

$$\begin{aligned}
T(\kappa_{\theta\theta}/\kappa_{rr} < 0) &= A_0 + B_1 \cos\theta \sin(m \ln r) \\
&\quad + D_1 \cos\theta \cos(m \ln r),
\end{aligned} \tag{14.6}$$

$$T\left(\kappa_{\theta\theta}/\kappa_{rr} > 0\right) = E_0 + F_1 r^n \cos\theta$$
$$+H_1 r^{-n} \cos\theta. \tag{14.7}$$

Therefore, we can obtain $T_m = E_0 + F_1 r \cos\theta + H_1 r^{-1} \cos\theta$ for an isotropic matrix. We set H_1 to be zero to ensure the external field undistorted. Then we can derive the effective permeability of the core-shell structure κ_e as

$$\kappa_e = m\kappa_{rr} \frac{\kappa_c + m\kappa_{rr} \tan\left(m \ln\sqrt{p}\right)}{m\kappa_{rr} - \kappa_c \tan\left(m \ln\sqrt{p}\right)}, \tag{14.8}$$

where $m = \sqrt{-\kappa_{\theta\theta}/\kappa_{rr}}$, and $p = (r_c/r_s)^2$ is the core fraction. As defined in this chapter, thermal chameleonlike metashells are characterized by the adaptive responses to inside objects. Namely, the effective thermal conductivity of the shell (κ_s) is always equal to that of the inside object,

$$\kappa_s = \kappa_c. \tag{14.9}$$

Based on the requirement of Eq. (14.9), the effective thermal conductivity of the core-shell structure (κ_e) must be

$$\kappa_e = \kappa_c. \tag{14.10}$$

Then, we should find some special relations to make Eq. (14.8) turn into Eq. (14.10). Fortunately, we find one

$$\sqrt{-\kappa_{\theta\theta}/\kappa_{rr}} \ln\sqrt{p} = -N^+\pi, \tag{14.11}$$

where N^+ ($=1, 2, 3, \ldots$) can be any positive integer. Clearly, the requirement of Eq. (14.10) is strictly satisfied with Eq. (14.11).

14.2.2 Isotropic Bilayer Schemes

We also discuss the properties of the core-shell-shell structure presented in Fig. 14.1b. We set the core with radius r_c and thermal conductivity κ_c, and the two shells with radius r_{s1} and r_{s2}, and thermal conductivities κ_{s1} and κ_{s2}, respectively. The effective thermal conductivity of the core-shell-shell structure κ_e can be expressed as

$$\kappa_e = \kappa_{s2} \frac{\kappa_{12} + \kappa_{s2} + (\kappa_{12} - \kappa_{s2})\, p_{12}}{\kappa_{12} + \kappa_{s2} - (\kappa_{12} - \kappa_{s2})\, p_{12}}, \tag{14.12}$$

where $p_{12} = (r_{s1}/r_{s2})^2$. κ_{12} is the effective thermal conductivity of the core plus the first shell, which can be calculated by

$$\kappa_{12} = \kappa_{s1} \frac{\kappa_c + \kappa_{s1} + (\kappa_c - \kappa_{s1}) \, p_c}{\kappa_c + \kappa_{s1} - (\kappa_c - \kappa_{s1}) \, p_c}, \tag{14.13}$$

where $p_c = (r_c/r_{s1})^2$.

We also find a special relation to make Eq. (14.12) turn into Eq. (14.10),

$$(\kappa_{s1} + \kappa_{s2})^2 + (p_c - p_{12})^2 = 0. \tag{14.14}$$

which gives that $\kappa_{s1} + \kappa_{s2} = 0$ and $p_c - p_{12} = 0$ should be simultaneously satisfied. Clearly, the requirement of Eq. (14.10) is strictly satisfied with Eq. (14.14).

So far, we have theoretically analyzed the thermal chameleonlike metashells in two dimensions, say Eq. (14.11) for monolayer schemes and Eq. (14.14) for bilayer schemes. The presence of thermal chameleonlike metashells can ensure that the effective thermal conductivity of the whole structure corresponds with the core, see Fig. 14.1c.

14.2.3 Three-Dimensional Counterpart of Anisotropic Monolayer Schemes

The theory for two-dimensional anisotropic monolayer schemes can be extended to three-dimensional schemes. Equation (14.1) in spherical coordinates is expressed as

$$\frac{1}{r} \frac{\partial}{\partial r} \left(r^2 \kappa_{rr} \frac{\partial T}{\partial r} \right) + \frac{1}{\sin \theta} \frac{\partial}{\partial \theta} \left(\sin \theta \kappa_{\theta\theta} \frac{\partial T}{r \partial \theta} \right) = 0. \tag{14.15}$$

The general solution of Eq. (14.15) is

$$T \left(\kappa_{\theta\theta} / \kappa_{rr} < -1/8 \right) = A_0 + B_0 r^{-1}$$
$$+ \sum_{i=1}^{\infty} r^{-1/2} \left[A_i \sin \left(s \ln r \right) + B_i \cos \left(s \ln r \right) \right]$$
$$\times P_i \left(\cos \theta \right), \tag{14.16}$$
$$T \left(-1/8 < \kappa_{\theta\theta} / \kappa_{rr} < 0 \right) = C_0 + D_0 r^{-1}$$
$$+ \sum_{i=1}^{j} \left(C_i r^{t_1} + D_i r^{t_2} \right) P_i \left(\cos \theta \right)$$
$$+ \sum_{i=j+1}^{\infty} r^{-1/2} \left[E_i \sin \left(s \ln r \right) + F_i \cos \left(s \ln r \right) \right]$$
$$\times P_i \left(\cos \theta \right), \tag{14.17}$$
$$T \left(0 \le \kappa_{\theta\theta} / \kappa_{rr} \right) = \sum_{i=0}^{\infty} \left(G_i r^{t_1} + H_i r^{t_2} \right)$$
$$\times P_i \left(\cos \theta \right), \tag{14.18}$$

where $s = \sqrt{-1/4 - i\,(i+1)\,\kappa_{\theta\theta}/\kappa_{rr}}$, $t_{1,2} = -1/2 \pm \sqrt{1/4 + i\,(i+1)\,\kappa_{\theta\theta}/\kappa_{rr}}$, i is the summation index, $j = \mathrm{INT}\left[-1/2 + \sqrt{1/4 - \kappa_{rr}/(4\kappa_{\theta\theta})}\right]$, and $\mathrm{INT}[\cdots]$ is the integral function with respect to $\cdots$. P_i is Legendre polynomials.

In fact, Eqs. (14.17) and (14.18) turn to be the same with similar boundary conditions of Eq. (14.5) because we only require to keep several terms of $i = 1$. Thus, Eqs. (14.16)–(14.18) can be simplified as

$$
\begin{aligned}
&T\left(\kappa_{\theta\theta}/\kappa_{rr} < -1/8\right) = A_0 \\
&+ r^{-1/2}\left[A_1 \sin\left(u \ln r\right) + B_1 \cos\left(u \ln r\right)\right]\cos\theta,
\end{aligned}
\tag{14.19}
$$

$$
\begin{aligned}
&T\left(\kappa_{\theta\theta}/\kappa_{rr} > -1/8\right) = G_0 \\
&+ \left(G_1 r^{v_1} + H_1 r^{v_2}\right)\cos\theta,
\end{aligned}
\tag{14.20}
$$

where $u = \sqrt{-1/4 - 2\kappa_{\theta\theta}/\kappa_{rr}}$, and $v_{1,2} = -1/2 \pm \sqrt{1/4 + 2\kappa_{\theta\theta}/\kappa_{rr}}$,

We set the core with radius r_c and scalar thermal conductivity κ_c, and the shell with radius r_s and tensorial thermal conductivity $\kappa_s = \mathrm{diag}\left(\kappa_{rr},\ \kappa_{\theta\theta},\ \kappa_{\varphi\varphi}\right)$ with $\kappa_{\theta\theta} = \kappa_{\varphi\varphi}$ for brevity. It should be noted that $\kappa_{\theta\theta}/\kappa_{rr} < -1/8$.

We can obtain $T_m = G_0 + \left(G_1 r + H_1 r^{-2}\right)\cos\theta$ for an isotropic matrix. We set H_1 to be zero to ensure the external field undistorted. Then we can derive the effective thermal conductivity of the core-shell structure κ_e in three dimensions as

$$
\kappa_e = \kappa_{rr} \frac{4u\kappa_c + \left[2\kappa_c + \left(1 + 4u^2\right)\kappa_{rr}\right]\tan\left(u \ln \sqrt[3]{p}\right)}{4u\kappa_{rr} - 2\left(2\kappa_c + \kappa_{rr}\right)\tan\left(u \ln \sqrt[3]{p}\right)},
\tag{14.21}
$$

where $p = \left(r_c/r_s\right)^3$ is the core fraction.

We also find a special relation to make Eq. (14.21) to satisfy the requirement of Eq. (14.10)

$$
\sqrt{-1/4 - 2\kappa_{\theta\theta}/\kappa_{rr}}\ \ln \sqrt[3]{p} = -N^+\pi,
\tag{14.22}
$$

where $N^+ (= 1,\ 2,\ 3,\ \ldots)$ can be any positive integer. Clearly, with Eq. (14.22), the requirement of Eq. (14.10) is perfectly satisfied. Therefore, thermal chameleonlike metashells can be achieved in three dimensions with anisotropic monolayer scheme.

14.2.4 Explanation for the Failure of Isotropic Bilayer Schemes in Three Dimensions

Then we consider the isotropic bilayer schemes in three dimensions. We set the core to have radius r_c and scalar thermal conductivity κ_c, and the two shells to have radii r_{s1} and r_{s2} and scalar thermal conductivities κ_{s1} and κ_{s2}. Then we can derive the effective thermal conductivity of the core-shell-shell structure κ_e as

$$
\kappa_e = \kappa_{s2} \frac{\kappa_{12} + 2\kappa_{s2} + 2\left(\kappa_{12} - \kappa_{s2}\right)p_{12}}{\kappa_{12} + 2\kappa_{s2} - \left(\kappa_{12} - \kappa_{s2}\right)p_{12}},
\tag{14.23}
$$

where $p_{12} = (r_{s1}/r_{s2})^3$. κ_{12} is the effective thermal conductivity of the core plus the first shell, which can be calculated as

$$\kappa_{12} = \kappa_{s1} \frac{\kappa_c + 2\kappa_{s1} + 2(\kappa_c - \kappa_{s1}) p_c}{\kappa_c + 2\kappa_{s1} - (\kappa_c - \kappa_{s1}) p_c}, \tag{14.24}$$

where $p_c = (r_c/r_{s1})^3$.

Although we find a special relation to make Eq. (14.23) turn into Eq. (14.10)

$$[(p_{12} - 2p_c + p_c p_{12}) \kappa_c - (p_c - 2p_{12} + p_c p_{12}) \kappa_{s1}]^2$$
$$+ (\kappa_{s1} + 2\kappa_{s2})^2 = 0, \tag{14.25}$$

Equation (14.25) is dependent on κ_c. This means that the chameleonlike behavior is dependent on the core property, which is not what we expect. Therefore, isotropic bilayer schemes fail in three dimensions.

14.3   Simulations of Thermal Chameleonlike Metashells

We perform finite-element simulations to validate the two proposed schemes with COMSOL Multiphysics (http://www.comsol.com/). To perform simulations, we set the thermal conductivities of the inside core and outside background to be the same. Then, we compare the results of thermal chameleonlike metashells (Fig. 14.2a, b), a normal shell (Fig. 14.2c), and a reference shell (Fig. 14.2d). Clearly, the same temperature profiles are obtained outside the thermal chameleonlike metashells (Fig. 14.2a, b) and the reference shell (Fig. 14.2d). Thus, thermal chameleonlike metashells do adaptively change their effective thermal conductivity according the nearby changes as expected. In contrast, the different background thermal profiles between the normal shell (Fig. 14.2c) and the reference shell (Fig. 14.2d) show that a normal shell does not possess the ability to change adaptively.

To validate the robustness, we change the thermal conductivities of the inside core and outside background, and keep those of the thermal chameleonlike metashells unchanged, to create a different condition; see Fig. 14.3. As expected, thermal chameleonlike metashells change their effective thermal conductivities adaptively according to the nearby changes. As a result, the temperature distributions outside the chameleonlike metashells (Fig. 14.3a, b) and the reference shell (Fig. 14.3d) are the same. However, that outside the normal shell (Fig. 14.3c) is different, which exhibits no adaptivity.

Finally, we consider an anisotropic case to show the capability of the thermal chameleonlike metashells. Similarly, we keep the thermal chameleonlike metashells unchanged, and change the thermal conductivities of the inside core and outside background to be anisotropic; see Fig. 14.4. The same conclusion can be obtained from the same background temperature distributions between Fig. 14.4a–d. Certainly, the normal shell fails again Fig. 14.4c.

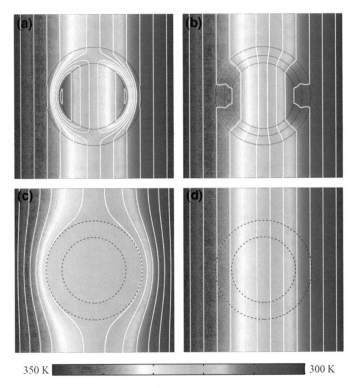

350 K ▨▨▨▨▨▨▨▨▨▨▨▨▨▨▨▨▨ 300 K

Fig. 14.2 The simulation box is $10 \times 10\,\mathrm{cm}^2$. The inner and outer radius in **a**, **c**, and **d** is 2 and 3 cm. The three radius in **b** are 2, $\sqrt{6}$, and 3 cm. The thermal conductivities of the thermal chameleonlike metashells are $\kappa_s = \mathrm{diag}\,(10,\ -600.33)\,\mathrm{Wm}^{-1}\mathrm{K}^{-1}$ for **a**, and $-10\,\mathrm{Wm}^{-1}\mathrm{K}^{-1}$ inner shell and 10 outer shell for **b**. Those of the inside core and outside background are $0.1\,\mathrm{Wm}^{-1}\mathrm{K}^{-1}$. That of the normal shell is $5\,\mathrm{Wm}^{-1}\mathrm{K}^{-1}$ throughout this chapter. Adapted from Ref. [33]

14.4   Discussion and Conclusion

We have proposed and confirmed the performance of two kinds of thermal chameleon-like metashells by both theoretical analyses and finite-element simulations. Our results have shown that the two schemes are exactly valid, indicating that they can work for various changes of nearby objects.

In this chapter, apparent negative thermal conductivities have been applied in our design. Although they do not occur in nature, they can be artificially realized by adding extra heat energy [30–32]. Thus, these two schemes have practical significance.

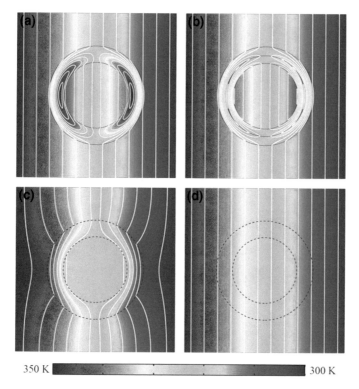

350 K ▬▬▬▬▬▬▬▬▬▬▬ 300 K

Fig. 14.3 All the parameters are the same as those for Fig. 14.2 except for the thermal conductivities of the inside core and outside background, say 100 $Wm^{-1}K^{-1}$. Adapted from Ref. [33]

In summary, we have presented two types of schemes to realize the thermal adaptive responses to the change of objects in the vicinity. Such schemes can work as multifunction materials to meet various requirements of thermal conductivities under different conditions. This chapter also provides guidance to other diffusive fields.

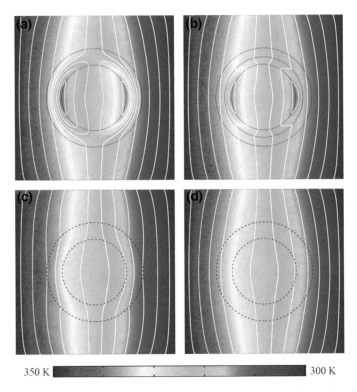

350 K ▮▮▮▮▮▮▮▮▮▮▮▮▮▮▮▮▮▮▮▮▮▮ 300 K

Fig. 14.4 All the parameters are the same as those for Fig. 14.2 except for the thermal conductivities of the inside core and outside background, say diag $(10, 20)$ Wm^{-1}K^{-1}. Adapted from Ref. [33]

References

1. Fan, C.Z., Gao, Y., Huang, J.P.: Shaped graded materials with an apparent negative thermal conductivity. Appl. Phys. Lett. **92**, 251907 (2008)
2. Narayana, S., Sato, Y.: Heat flux manipulation with engineered thermal materials. Phys. Rev. Lett. **108**, 214303 (2012)
3. Schittny, R., Kadic, M., Guenneau, S., Wegener, M.: Experiments on transformation thermodynamics: molding the flow of heat. Phys. Rev. Lett. **110**, 195901 (2013)
4. Xu, H.Y., Shi, X.H., Gao, F., Sun, H.D., Zhang, B.L.: Ultrathin three-dimensional thermal cloak. Phys. Rev. Lett. **112**, 054301 (2014)
5. Han, T.C., Bai, X., Gao, D.L., Thong, J.T.L., Li, B.W., Qiu, C.-W.: Experimental demonstration of a bilayer thermal cloak. Phys. Rev. Lett. **112**, 054302 (2014)
6. Ma, Y.G., Liu, Y.C., Raza, M., Wang, Y.D., He, S.L.: Experimental demonstration of a multiphysics cloak: manipulating heat flux and electric current simultaneously. Phys. Rev. Lett. **113**, 205501 (2014)
7. Shen, X.Y., Huang, J.P.: Thermally hiding an object inside a cloak with feeling. Int. J. Heat Mass Transfer **78**, 1 (2014)
8. Li, Y., Zhu, K.J., Peng, Y.G., Li, W., Yang, T.Z., Xu, H.X., Chen, H., Zhu, X.F., Fan, S.H., Qiu, C.W.: Thermal meta-device in analogue of zero-index photonics. Nat. Mater. **18**, 48–54 (2019)
9. Han, T.C., Yang, P., Li, Y., Lei, D.Y., Li, B.W., Hippalgaonkar, K., Qiu, W.: Full-parameter omnidirectional thermal metadevices of anisotropic geometry. Adv. Mater. **30**, 1804019 (2018)

10. Guo, J., Qu, Z.G.: Thermal cloak with adaptive heat source to proactively manipulate temperature field in heat conduction process. Int. J. Heat Mass Transfer **127**, 1212 (2018)
11. He, X., Wu, L.Z.: Thermal transparency with the concept of neutral inclusion. Phys. Rev. E **88**, 033201 (2013)
12. Zeng, L.W., Song, R.X.: Experimental observation of heat transparency. Appl. Phys. Lett. **104**, 201905 (2014)
13. Yang, T.Z., Bai, X., Gao, D.L., Wu, L.Z., Li, B.W., Thong, J.T.L., Qiu, C.W.: Invisible sensors: simultaneous sensing and camouflaging in multiphysical fields. Adv. Mater. **27**, 7752–7758 (2015)
14. Xu, L.J., Yang, S., Huang, J.P.: Thermal transparency induced by periodic interparticle interaction. Phys. Rev. Appl. **11**, 034056 (2019)
15. Vemuri, K.P., Bandaru, P.R.: Anomalous refraction of heat flux in thermal metamaterials. Appl. Phys. Lett. **104**, 083901 (2014)
16. Yang, T.Z., Vemuri, K.P., Bandaru, P.R.: Experimental evidence for the bending of heat flux in a thermal metamaterial. Appl. Phys. Lett. **105**, 083908 (2014)
17. Vemuri, K.P., Canbazoglu, F.M., Bandaru, P.R.: Guiding conductive heat flux through thermal metamaterials. Appl. Phys. Lett. **105**, 193904 (2014)
18. Xu, G.Q., Zhang, H.C., Jin, Y., Li, S., Li, Y.: Control and design heat flux bending in thermal devices with transformation optics. Opt. Express **25**, A419 (2017)
19. Han, T.C., Bai, X., Thong, J.T.L., Li, B.W., Qiu, C.-W.: Full control and manipulation of heat signatures: cloaking, camouflage and thermal metamaterials. Adv. Mat. **26**, 1731–1734 (2014)
20. He, X., Wu, L.Z.: Illusion thermodynamics: a camouflage technique changing an object into another one with arbitrary cross section. Appl. Phys. Lett. **105**, 221904 (2014)
21. Yang, T.Z., Su, Y.S., Xu, W.K., Yang, X.D.: Transient thermal camouflage and heat signature control. Appl. Phys. Lett. **109**, 121905 (2016)
22. Xu, G.Q., Zhang, H.C., Zou, Q., Jin, Y., Xie, M.: Forecast of thermal harvesting performance under multi-parameter interaction with response surface methodology. Int. J. Heat Mass Transfer **115**, 682 (2017)
23. Hu, R., Zhou, S.L., Li, Y., Lei, D.Y., Luo, X.B., Qiu, C.W.: Illusion thermotics. Adv. Mater. **30**, 1707237 (2018)
24. Zhou, S.L., Hu, R., Luo, X.B.: Thermal illusion with twinborn-like heat signatures. Int. J. Heat Mass Transfer **127**, 607 (2018)
25. Li, Y., Bai, X., Yang, T.Z., Luo, H., Qiu, C.W.: Structured thermal surface for radiative camouflage. Nat. Commun. **9**, 273 (2018)
26. Yang, S., Xu, L.J., Wang, R.Z., Huang, J.P.: Full control of heat transfer in single-particle structural materials. Appl. Phys. Lett. **111**, 121908 (2017)
27. Xu, L.J., Yang, S., Huang, J.P.: Passive metashells with adaptive thermal conductivities: chameleonlike behavior and its origin. Phys. Rev. Appl. **11**, 054071 (2019)
28. Xu, L.J., Huang, J.P.: Magnetostatic chameleonlike metashells with negative permeabilities. EPL (Europhys. Lett.) **125**, 64001 (2019)
29. Xu, L.J., Huang, J.P.: Electrostatic chameleons: theory of intelligent metashells with adaptive response to inside objects. Eur. Phys. J. B **92**, 53 (2019)
30. Yang, S., Xu, L.J., Huang, J.P.: Thermal magnifier and external cloak in ternary component structure. J. Appl. Phys. **125**, 055103 (2019)
31. Yang, S., Xu, L.J., Huang, J.P.: Metathermotics: nonlinear thermal responses of core-shell metamaterials. Phys. Rev. E **99**, 042144 (2019)
32. Yang, S., Xu, L.J., Huang, J.P.: Intelligence thermotics: correlated self-fixing behavior of thermal metamaterials. EPL (Europhys. Lett.) **126**, 54001 (2019)
33. Yang, S., Xu, L.J., Huang, J.P.: Two exact schemes to realize thermal chameleonlike metashells. EPL (Europhys. Lett.) (in press) (2019)

Chapter 15
Theory for Anisotropic Core and Isotropic Shell: Isothermal Rotation

Abstract Architected structures have aroused widespread research interest for they possess unique properties in mechanics. However, a fundamental theory has not been established to understand their thermal properties. This chapter describes a theoretical framework in thermotics to predict thermal properties of architected structures. Then, by experiment and simulation, we show its applications in the field of heat management. By assembling two radically different materials, we design two types of Janus structures. The different rotation degrees of the Janus structures can flexibly control the switch between different functions, such as, from partial concentration to uniform concentration and from rotation to concentration. These functions are realized in the structure made of heterogeneous core plus homogeneous shell, which is contrary to the existing structures made of homogeneous core plus heterogeneous shell designed by the theory of transformation thermotics. This chapter lays a theoretical foundation in thermotics for further research on heterogeneously architected structures, and it proposes the concept of thermal Janus structures for flexible heat control, which may open an avenue for intelligent thermal metamaterials.

Keywords Architected structure · Janus structure · Intelligent thermal metamaterial · Rotation · Concentration

15.1 Opening Remarks

Heat manipulation is of growing significance due to the universality of heat energy. Fortunately, with delicate design of architected structures, many unique thermal phenomena have been realized, which include cloaking or camouflage [1–15], concentration [4, 9, 16, 17], rotation [4, 18], transparency [19], and their combinations [20–25]. These functions are mostly based on the structure of a coated core, which is generally featured by a homogeneous core and a heterogeneous (architected) shell according to the theory of transformation thermotics [1–3]. Even so, heterogeneously architected structures have aroused less attention in thermotics than in mechanics (e.g., see Ref. [26] and references therein). This situation mainly results from the lack of enough fundamental thermal theories for handling heterogeneously architected structures.

© Springer Nature Singapore Pte Ltd. 2020
J.-P. Huang, *Theoretical Thermotics*,
https://doi.org/10.1007/978-981-15-2301-4_15

As a meaningful attempt, here we introduce the structure of a coated core, which is, however, composed of a heterogeneously architected core and a homogeneous shell. A theoretical framework will be established to predict the effective thermal response of the coated core.

In addition, another challenge is to realize multi-functions in a single field. Compared with multi-functions in multi-fields [20–24], there is only one adjustable parameter in a single field (such as thermal conductivity in the single thermal field), so the realization of multi-functions in a single field becomes relatively more difficult. Although Shen et al. [25] have realized a type of cloak-concentrator in thermotics by tailoring the temperature-dependent effect of thermal conductivities, different mechanisms for multi-functions still remain to be investigated.

To overcome this challenge, here we further propose a concept of thermal Janus core on the basis of the aforementioned theoretical framework, which can be seen as a typical kind of heterogeneously architected core. Generally speaking, such a Janus core is composed of two radically different materials, which has been widely studied in soft matter (e.g., see Refs. [27–29]). Then we experimentally fabricate two samples to realize the flexible control of isotherm concentration by rotating the core. We further propose another concept of generalized thermal Janus core, which is also composed of two radically different materials, but with more flexible structures. As a result, the switch between thermal rotation and concentration can be achieved by rotating the core.

15.2 Theory

Let us start by investigating the effective thermal conductivity of a two-dimensional coated core with circular shape; see Fig. 15.1a. The thermal property of the core is represented by a heterogeneous, anisotropic and diagonal thermal conductivity tensor $\bar{\kappa}_c (x, y)$,

$$\bar{\kappa}_c (x, y) = \begin{bmatrix} \kappa_{xx} (x, y) & 0 \\ 0 & \kappa_{yy} (x, y) \end{bmatrix}. \tag{15.1}$$

We then write down the boundary condition [heat flux $\boldsymbol{J}_b (x, y)$] on Boundary I [see Fig. 15.1b] in Cartesian coordinates according to Fourier's law,

$$\begin{aligned} \boldsymbol{J}_b (x, y) &= \begin{bmatrix} J_x (x, y) \\ J_y (x, y) \end{bmatrix} = -\bar{\kappa}_c (x, y) \, \nabla T_b (x, y) \\ &= -\begin{bmatrix} \kappa_{xx} (x, y) \, \partial T_b (x, y) / \partial x \\ \kappa_{yy} (x, y) \, \partial T_b (x, y) / \partial y \end{bmatrix}, \end{aligned} \tag{15.2}$$

where $T_b (x, y)$ is the temperature distribution on Boundary I (see Fig. 15.1b). We rewrite the heat flux $\boldsymbol{J}_b (x, y)$ in polar coordinates for the convenience of discussion,

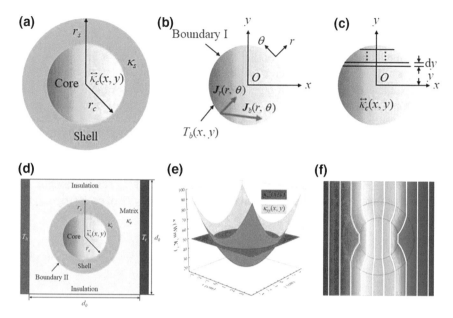

Fig. 15.1 Schematic diagrams (**a–c**) and simulation settings (**d, e**) of a coated core. **a** For structures and parameters, **b** for boundary conditions (heat flux and temperature), and **c** for calculating method. **d** for simulation box, and **e** for thermal conductivities of the core. **f** Shows the simulation result with $\bar{\kappa}_c = \text{diag}\left[50,\ 20 + 30000\left(x^2 + y^2\right)\right]$, $\kappa_s = 400$, and $\kappa_e = 242\,\text{Wm}^{-1}\text{K}^{-1}$. Parameters: $T_h = 313\,\text{K}$, $T_c = 273\,\text{K}$, $d_0 = 20\,\text{cm}$, $r_c = 3.6\,\text{cm}$, and $r_s = 6.4\,\text{cm}$. White lines represent isotherms. Adapted from Ref. [32]

$$\boldsymbol{J}_b\left(r,\ \theta\right) = \begin{bmatrix} J_r\left(r,\ \theta\right) \\ J_\theta\left(r,\ \theta\right) \end{bmatrix} = \begin{bmatrix} J_x\left(x,\ y\right)\cos\theta + J_y\left(x,\ y\right)\sin\theta \\ -J_x\left(x,\ y\right)\sin\theta + J_y\left(x,\ y\right)\cos\theta \end{bmatrix}, \quad (15.3)$$

where $x = r\cos\theta$, $y = r\sin\theta$.

Now we pay attention to the radial component $J_r\left(r,\ \theta\right)$ (see Fig. 15.1b), for heat flux is always conservative along the radial direction of Boundary I,

$$J_r\left(r,\ \theta\right) = -\kappa_{xx}\left(x,\ y\right)\cos\theta\partial T_b\left(x,\ y\right)/\partial x - \kappa_{yy}\left(x,\ y\right)\sin\theta\partial T_b\left(x,\ y\right)/\partial y. \quad (15.4)$$

For further discussion, we have to make an approximation for boundary condition [temperature $T_b\left(x,\ y\right)$]. Namely, when the direction of external thermal field is along x axis and $\kappa_{xx}\left(x,\ y\right)$ is a constant, the temperature distribution on Boundary I is uniform,

$$T_b\left(x,\ y\right) = Ax + B, \quad (15.5)$$

where A and B are two constants. We suppose that Eq. (15.5) holds approximately when $\kappa_{xx}\left(x,\ y\right)$ experiences a small variation. Therefore, the radial component of heat flux Eq. (15.4) can be simplified as

$$J_r\,(r,\,\theta) = -\kappa_{xx}\,(x,\,y)\,A\cos\theta. \tag{15.6}$$

The most important feature of Eq. (15.6) is that $J_r\,(r,\,\theta)$ is independent of $\kappa_{yy}\,(x,\,y)$, which means that the effective thermal conductivity of the coated core is independent of $\kappa_{yy}\,(x,\,y)$. In other words, we only need to calculate the effective thermal conductivity of $\kappa_{xx}\,(x,\,y)$ (denoted by scalar κ_c), and then the effective thermal conductivity of the coated core can be obtained. Therefore, as long as the variation of $\kappa_{xx}\,(x,\,y)$ is small enough, Eq. (15.5) is rational and contributing.

Now we are in a position to calculate the effective thermal conductivity of $\kappa_{xx}\,(x,\,y)$, namely κ_c. The concrete approximation method is as follows.

We first separate the core into strips whose width dy is small enough; see Fig. 15.1c. Then we calculate the effective thermal conductivity of each strip $\kappa_{xx}\,(y)$ through series connection,

$$\kappa_{xx}\,(y) = \frac{2\sqrt{r_c^2 - y^2}}{\displaystyle\int_{-\sqrt{r_c^2-y^2}}^{\sqrt{r_c^2-y^2}} \frac{dx}{\kappa_{xx}\,(x,\,y)}}, \tag{15.7}$$

where r_c is the radius of the core.

We further calculate the effective thermal conductivity of these strips through parallel connection,

$$\kappa_c = \int_{-r_c}^{r_c} \frac{2\kappa_{xx}\,(y)\,\sqrt{r_c^2 - y^2}\,dy}{\pi r_c^2}. \tag{15.8}$$

We finally calculate the effective thermal conductivity of the coated core κ_e through the single-particle effective medium theory, namely Eq. (11) in Ref. [30],

$$\kappa_e = \kappa_s \frac{\kappa_c + \kappa_s + (\kappa_c - \kappa_s)p}{\kappa_c + \kappa_s - (\kappa_c - \kappa_s)p}, \tag{15.9}$$

where κ_s is the thermal conductivity of the shell, and $p = (r_c/r_s)^2$ is the area fraction of the core.

The final results Eqs. (15.7)–(15.9) are independent of $\kappa_{yy}\,(x,\,y)$ under the approximation condition of Eq. (15.5). To be mentioned, the influence of the conductivity variation on the accuracy of Eqs. (15.7)–(15.9) is analyzed in section III via finite-element simulations.

Similarly, we can derive the effective thermal conductivity for a three-dimensional coated core with spherical shape, as shown in the Supplementary Proof.

15.3 Simulation

We simulate the two-dimensional coated core (see Fig. 15.1d) to observe the performance of Eqs. (15.7)–(15.9). For arbitrary $\kappa_{xx}(x, y)$, well-defined (or strict) effective thermal conductivity of a coated core probably does not exist. Therefore, we define a relative error Δ to indirectly examine the performance of Eqs. (15.7)–(15.9) on predicting the effective thermal conductivity,

$$\Delta = \frac{\oint_\Sigma |\nabla T - \nabla T_0| \, dl}{|\nabla T_0| \oint_\Sigma dl}, \tag{15.10}$$

where ∇T_0 is the external thermal field, $|\nabla T_0| = (T_h - T_c)/d_0$, and the integrate boundary Σ is the outer periphery of the coated core (denoted by Boundary II); see Fig. 15.1d. Clearly, better performance of Eqs. (15.7)–(15.9) corresponds to less influence of the coated core on the matrix, which is represented by a smaller value of Δ. For our purpose, we calculate Eq. (15.10) by using the finite-element simulation based on the commercial software COMSOL Multiphysics (http://www.comsol.com/).

We have mentioned that when $\kappa_{xx}(x, y)$ is a constant, Eq. (15.5) is strictly satisfied and κ_e is independent of $\kappa_{yy}(x, y)$. To validate the statement, we set $\kappa_{xx}(x, y) = 50\,\mathrm{Wm^{-1}K^{-1}}$, and arbitrary $\kappa_{yy}(x, y) = 20 + 30000\,(x^2 + y^2)$ $\mathrm{Wm^{-1}K^{-1}}$; see Fig. 15.1e. The thermal conductivity of the shell κ_s is set to be $400\,\mathrm{Wm^{-1}K^{-1}}$, and the effective thermal conductivity of the coated core is calculated from Eqs. (15.7)–(15.9). We find that the temperature gradient in the core is uniform; see Fig. 15.1f, and obtain $\Delta = 0$ according to Eq. (15.10). The results indicate that when $\kappa_{xx}(x, y)$ is a constant, Eqs. (15.7)–(15.9) can be reduced to account for the known case of uniform thermal conductivities.

Now we discuss a position-dependent $\kappa_{xx}(x, y)$. Our theory [Eqs. (15.7)–(15.9)] does not consider the effect of $\kappa_{yy}(x, y)$, so we set $\kappa_{yy}(x, y) = \kappa_{xx}(x, y)$ without loss of generality. We choose two typical functions: $F(w) = 20 + |100w|^C$ and $G(w) = 20 + 60/(1 + e^{-100Dw})$, which are respectively even-symmetric and odd-symmetric; see Fig. 15.2a, d. To distinguish the contribution, we respectively set $\kappa_{xx}(x, y) = F(x)$, $\kappa_{xx}(x, y) = F(y)$, $\kappa_{xx}(x, y) = G(x)$, and $\kappa_{xx}(x, y) = G(y)$. The effective thermal conductivities and relative errors are respectively shown in Fig. 15.2b, e and c, f. The maximum relative error is below 1.9%, which shows the ability of our theory to predict the effective thermal conductivities. In $F(w)$ and $G(w)$, C can reflect the variation amplitude, and $\ln D$ can reflect the variation speed. Comparing Fig. 15.2c, f, we find that the relative errors increase as the variation amplitude (C) and variation speed ($\ln D$) increase. In addition, the relative errors of even-symmetric distribution [$F(w)$] of thermal conductivities are smaller than those of odd-symmetric distribution [$G(w)$].

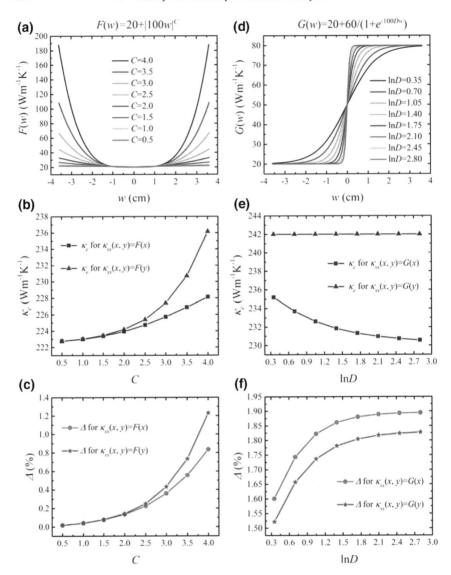

Fig. 15.2 Simulation results of a coated core. **a**, **d** are two typical functions which will be set as the thermal conductivities of the core. **b**, **e** are the predicted effective thermal conductivities of the coated core with Eqs. (15.7)–(15.9). **c**, **f** are the relative errors of the predicted thermal conductivities with Eq. (15.10). Other simulation settings are same as those for Fig. 15.2d. Adapted from Ref. [32]

15.4 Application: Experiment and Simulation

15.4.1 Thermal Janus Core

The above theories and simulations pave the way to propose a concept of thermal Janus core which is composed of two radically different materials (material A and material B); see Fig. 15.3a. Here we apply such thermal Janus core to realize the manipulation of isotherm concentration; see Fig. 15.3b, c. We can observe the partial concentration of the isotherms in Fig. 15.3b. The isotherms become concentrated in the left part of the core, while sparse in the right. Then we anticlockwise rotate the thermal Janus core by 90°, and we observe the uniform concentration of the isotherms in Fig. 15.3c. Concrete concentration ratio is shown in Fig. 15.3j. To ensure that the rotation will not disturb the external field, the thermal conductivity of the core is required to satisfy

$$(\kappa_c)_0 = (\kappa_c)_{90}, \tag{15.11}$$

where the subscripts 0 and 90 respectively represent the anticlockwise rotation angle of the core, and they can be calculated from Eqs. (15.7)–(15.9). When calculating $(\kappa_c)_{90}$, the $\kappa_{yy}(x, y)$ in Eq. (15.1) actually works, and hence $\kappa_{xx}(x, y)$ in Eqs. (15.7)–(15.9) should be replaced by $\kappa_{yy}(x, y)$.

In the mean time, we also conduct experiments for demonstration. We drill different holes on a copper plate to design practical structures of the core, and fabricate two samples with laser engraving; see Fig. 15.3d, e, f. Detailed parameters of the two experimental samples are designed according to the periodic-particle effective medium theory, namely Eq. (4) in Ref. [31], which are shown in Fig. 15.3g. We use water baths to act as hot or cold sources, and an infrared camera Flir E60 to detect the thermal profile. The measurements are conducted at standard atmosphere pressure and room temperature. The measured and simulated results based on the two samples are respectively shown in Fig. 15.3h, i and k, l. Clearly, good agreement with our theory has come to appear.

Then we rotate the core by 22.5, 45, 67.5° to observe the switch progress from partial concentration to uniform concentration: Fig. 15.4a–c for anticlockwise rotation, Fig. 15.4d–f for clockwise rotation. Owing to Eq. (15.11), the external field is not disturbed as expected.

15.4.2 Generalized Thermal Janus Core

We further propose another concept of generalized thermal Janus core which is also composed of two radically different materials (material A and material B), but with more flexible structures; see Fig. 15.5a. Here we apply such generalized thermal Janus core to realize the switch between concentration and rotation; see Fig. 15.5b, c. Figure 15.5b shows that the heat flux in the core has a rotation of

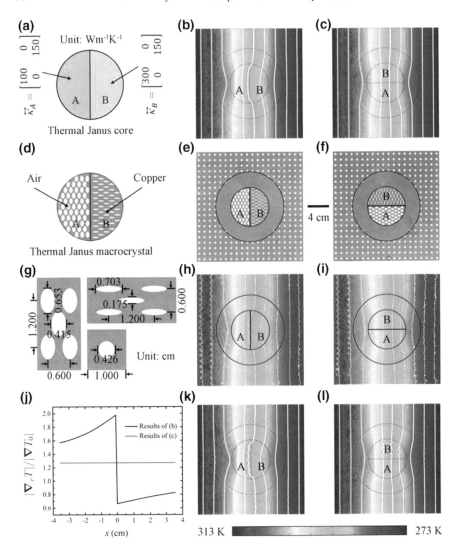

Fig. 15.3 Thermal Janus core. **a**, **d** for ideal thermal Janus core and practical thermal Janus macro-crystal. **b** is the simulated result of partial concentration (left part of the core). **c** is the simulated results of uniform concentration where the core is anticlockwise rotated by 90°. **e**, **f** are two fabricated samples whose measured and simulated results are respectively shown in **h**, **i** and **k**, **l**. The detailed parameters are displayed in **g**. **j** shows the concentration ratio in the core along x axis (where $y = 0$). Parameters: $\bar{\kappa}_A = \mathrm{diag}\,[100,\ 150]\,\mathrm{W\,m^{-1}\,K^{-1}}$, $\bar{\kappa}_B = \mathrm{diag}\,[300,\ 150]\,\mathrm{W\,m^{-1}\,K^{-1}}$, Copper: $400\,\mathrm{W\,m^{-1}\,K^{-1}}$, Air: $0.026\,\mathrm{W\,m^{-1}\,K^{-1}}$, and other parameters are same as those for Fig. 15.2d. Adapted from Ref. [32]

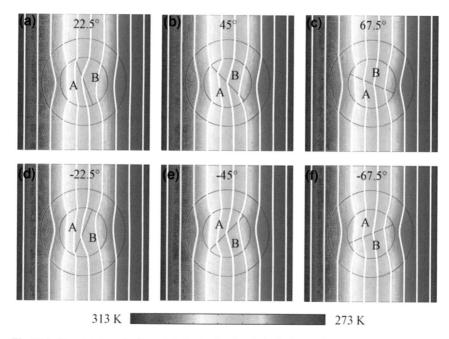

313 K ▮▮▮▮▮▮▮▮▮▮▮▮▮▮ 273 K

Fig. 15.4 Simulated results for anticlockwise (**a–c**) and clockwise (**d–f**) rotating the core by 22.5, 45, 67.5°. Other parameters are same as those for Fig. 15.3b. Adapted from Ref. [32]

about 62°, when compared with that in the matrix. Then we anticlockwise rotate the generalized thermal Janus core by 90°, and we can observe the switch from rotation to concentration in Fig. 15.5c. Incidentally, when designing the generalized thermal Janus core, Eq. (15.11) should be satisfied as well.

To experimentally demonstrate the validity of Fig. 15.5b, c, one might also drill different holes on a copper and plumbum plate to design practical structures of the core, and then design two samples; see Fig. 15.5d, e, f. Detailed parameters of the two samples are shown in Fig. 15.5g. The simulated results based on the two samples are shown in Fig. 15.5h, i, which agree well with our theoretical prediction as shown in Fig. 15.5b, c.

In addition, when we rotate the core by 22.5, 45, and 67.5°, we can observe the switch progress from rotation to concentration: Fig. 15.6a–c for anticlockwise rotating the core and Fig. 15.6d–f for clockwise rotating the core. We also adjust the parameters of h_1, h_2, w_1 and w_2 to observe the change in the rotation of heat flux; see Fig. 15.6g–i. To be mentioned, these parameters may affect κ_e according to Eqs. (15.7)–(15.9), but their influence is small enough to be neglected. The rotation degrees of heat flux in Fig. 15.6a–i are about 26, –20, –18, 55, 38, 18, 68, 60, and 48°. The results show that the generalized thermal Janus core has a flexible control of heat flux rotation while keeping the external field undisturbed.

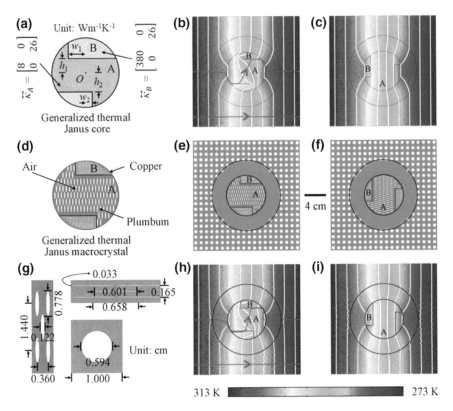

Fig. 15.5 Generalized thermal Janus core. Purple lines in **b**, **c**, **h**, **i** represent the flow of heat. **a**, **d** for ideal generalized thermal Janus core and practical generalized thermal Janus macrocrystal, respectively. **b** is the simulated result of heat flux rotation. **c** is the simulated results of concentration where the core is anticlockwise rotated by $90°$. **e**, **f** are two designed samples whose simulated results are shown in **h**, **i**. The detailed parameters are displayed in **g**. Parameters: $\bar{\kappa}_A = \text{diag}[8, 26]\,\text{Wm}^{-1}\text{K}^{-1}$, $\bar{\kappa}_B = \text{diag}[380, 26]\,\text{Wm}^{-1}\text{K}^{-1}$, Plumbum: $35\,\text{Wm}^{-1}\text{K}^{-1}$, $h_1 = h_2 = 2.12\,\text{cm}$, $w_1 = w_2 = 1.11\,\text{cm}$, and other parameters are same as those for Fig. 15.2d. Adapted from Ref. [32]

15.5   Conclusion

In summary, we have presented a theoretical framework to predict the effective thermal conductivity of a coated core with a heterogeneously architected core plus a homogeneous shell, which differs from the structure of homogeneous core plus a heterogeneously architected shell as extensively adopted for thermal rotation and concentration according to the theory of transformation thermotics. Based on the theory, we have proposed two kinds of Janus structures, which enable flexible heat manipulation for thermal rotation and concentration. Our theory has been confirmed by numerical simulations and our design of Janus structures has been validated by

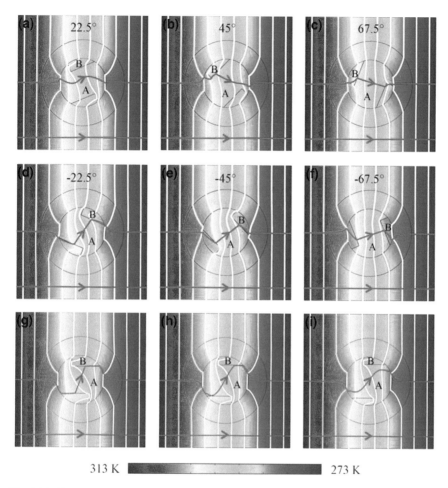

Fig. 15.6 Simulated results for other parameters. Anticlockwise (**a–c**) and clockwise (**d–f**) rotating the core by 22.5, 45, 67.5 °. **g–i** for changing the parameters of h_1, h_2, w_1 and w_2. Parameters: $h_1 = h_2 = 2.50$ cm and $w_1 = w_2 = 0.75$ cm for **g**, $h_1 = h_2 = 2.31$ cm and $w_1 = w_2 = 0.99$ cm for **h**, $h_1 = h_2 = 2.01$ cm and $w_1 = w_2 = 1.35$ cm for **i**, and other parameters are same as those for Fig. 15.5b. Adapted from Ref. [32]

both experiment and simulation. This chapter has not only potential applications in heat management, but also instructive meanings for exploring novel thermal phenomena of thermal metamaterials.

15.6   Supplementary Proof

Here we extend the theory from two dimensions (circular shape) to three dimensions (spherical shape). Then, the thermal property of the core can be represented by a heterogeneous, anisotropic and diagonal thermal conductivity tensor $\bar{\kappa}_c (x, y, z)$,

$$
\bar{\kappa}_c (x, y, z) = \begin{bmatrix} \kappa_{xx} (x, y, z) & 0 & 0 \\ 0 & \kappa_{yy} (x, y, z) & 0 \\ 0 & 0 & \kappa_{zz} (x, y, z) \end{bmatrix}. \tag{15.12}
$$

By imitating the two-dimensional results, Eqs. (15.7)–(15.9), we separate the core into sticks and calculate the effective thermal conductivity of each stick $\kappa_{xx} (y, z)$ through series connection pattern,

$$
\kappa_{xx} (y, z) = \frac{2\sqrt{r_c^2 - y^2 - z^2}}{\int_{-\sqrt{r_c^2 - y^2 - z^2}}^{\sqrt{r_c^2 - y^2 - z^2}} \frac{dx}{\kappa_{xx} (x, y, z)}}, \tag{15.13}
$$

where r_c is the radius of the core.

We further calculate the effective thermal conductivity of these strips through parallel connection pattern,

$$
\kappa_c = \int_{-r_c}^{r_c} \frac{2\kappa_{xx} (y, z) \sqrt{r_c^2 - y^2 - z^2}\, dy\, dz}{(4/3)\, \pi r_c^3}. \tag{15.14}
$$

We finally calculate the effective thermal conductivity of the coated core κ_e through the single-particle effective medium theory, namely Eq. (11) in Ref. [30],

$$
\kappa_e = \kappa_s \frac{\kappa_c + 2\kappa_s + 2(\kappa_c - \kappa_s)p}{\kappa_c + 2\kappa_s - (\kappa_c - \kappa_s)p}, \tag{15.15}
$$

where κ_s is the thermal conductivity of the shell, and $p = (r_c/r_s)^3$ is the volume fraction of the core.

15.7   Exercises and Solutions

Exercises

1. Consider a two-dimensional Janus core with radius r_c, whose left part is with tensorial thermal conductivity $\bar{\kappa}_A = \mathrm{diag}\,[100,\ 150]\,\mathrm{Wm^{-1}K^{-1}}$, and the right part is with that $\bar{\kappa}_B = \mathrm{diag}\,[300,\ 150]\,\mathrm{Wm^{-1}K^{-1}}$, which are expressed in Cartesian coordinates. Please calculate the approximate effective thermal conductivity along the x axis.

Solutions

1. **Solution:** $\kappa_{xx}(y) = 2\sqrt{r_c^2 - y^2} / \int_{-\sqrt{r_c^2-y^2}}^{\sqrt{r_c^2-y^2}} \dfrac{dx}{\kappa_{xx}(x,\ y)} = \dfrac{2}{1/100 + 1/300} =$

$150\,\mathrm{Wm^{-1}K^{-1}}$, and then $\kappa_c = \displaystyle\int_{-r_c}^{r_c} \dfrac{2\kappa_{xx}(y)\sqrt{r_c^2 - y^2}\,dy}{\pi r_c^2} = 150\,\mathrm{Wm^{-1}K^{-1}}$.

References

1. Fan, C.Z., Gao, Y., Huang, J.P.: Shaped graded materials with an apparent negative thermal conductivity. Appl. Phys. Lett. **92**, 251907 (2008)
2. Chen, T.Y., Weng, C.N., Chen, J.S.: Cloak for curvilinearly anisotropic media in conduction. Appl. Phys. Lett. **93**, 114103 (2008)
3. Guenneau, S., Amra, C., Veynante, D.: Transformation thermodynamics: cloaking and concentrating heat flux. Opt. Express **20**, 8207–8218 (2012)
4. Narayana, S., Sato, Y.: Heat flux manipulation with engineered thermal materials. Phys. Rev. Lett. **108**, 214303 (2012)
5. Schittny, R., Kadic, M., Guenneau, S., Wegener, M.: Experiments on transformation thermodynamics: molding the flow of heat. Phys. Rev. Lett. **110**, 195901 (2013)
6. Xu, H.Y., Shi, X.H., Gao, F., Sun, H.D., Zhang, B.L.: Ultrathin three-dimensional thermal cloak. Phys. Rev. Lett. **112**, 054301 (2014)
7. Han, T.C., Bai, X., Gao, D.L., Thong, J.T.L., Li, B.W., Qiu, C.-W.: Experimental demonstration of a bilayer thermal cloak. Phys. Rev. Lett. **112**, 054302 (2014)
8. Han, T.C., Bai, X., Thong, J.T.L., Li, B.W., Qiu, C.-W.: Full control and manipulation of heat signatures: cloaking, camouflage and thermal metamaterials. Adv. Mat. **26**, 1731–1734 (2014)
9. Chen, T.Y., Weng, C.N., Tsai, Y.L.: Materials with constant anisotropic conductivity as a thermal cloak or concentrator. J. Appl. Phys. **117**, 054904 (2015)
10. Yang, T.Z., Su, Y.S., Xu, W.K., Yang, X.D.: Transient thermal camouflage and heat signature control. Appl. Phys. Lett. **109**, 121905 (2016)
11. Hu, R., Zhou, S.L., Li, Y., Lei, D.Y., Luo, X.B., Qiu, C.W.: Illusion thermotics. Adv. Mater. **30**, 1707237 (2018)
12. Zhou, S.L., Hu, R., Luo, X.B.: Thermal illusion with twinborn-like heat signatures. Int. J. Heat Mass Tran. **127**, 607 (2018)
13. Wang, R.Z., Xu, L.J., Ji, Q., Huang, J.P.: A thermal theory for unifying and designing transparency, concentrating and cloaking. J. Appl. Phys. **123**, 115117 (2018)
14. Xu, L.J., Wang, R.Z., Huang, J.P.: Camouflage thermotics: a cavity without disturbing heat signatures outside. J. Appl. Phys. **123**, 245111 (2018)
15. Dai, G.L., Shang, J., Huang, J.P.: Theory of transformation thermal convection for creeping flow in porous media: cloaking, concentrating, and camouflage. Phys. Rev. E **97**, 022129 (2018)
16. Kapadia, R.S., Bandaru, P.R.: Heat flux concentration through polymeric thermal lenses. Appl. Phys. Lett. **105**, 233903 (2014)
17. Li, Y., Shen, X.Y., Huang, J.P., Ni, Y.S.: Temperature-dependent transformation thermotics for unsteady states: switchable concentrator for transient heat flow. Phys. Lett. A **380**, 1641 (2016)
18. Guenneau, S., Amra, C.: Anisotropic conductivity rotates heat fluxes in transient regimes. Opt. Express **21**, 6578–6583 (2013)
19. He, X., Wu, L.Z.: Thermal transparency with the concept of neutral inclusion. Phys. Rev. E **88**, 033201 (2013)
20. Li, J.Y., Gao, Y., Huang, J.P.: A bifunctional cloak using transformation media. J. Appl. Phys. **108**, 074504 (2010)

21. Ma, Y.G., Liu, Y.C., Raza, M., Wang, Y.D., He, S.L.: Experimental demonstration of a multiphysics cloak: manipulating heat flux and electric current simultaneously. Phys. Rev. Lett. **113**, 205501 (2014)
22. Moccia, M., Castaldi, G., Savo, S., Sato, Y., Galdi, V.: Independent manipulation of heat and electrical current via bifunctional metamaterials. Phys. Rev. X **4**, 021025 (2014)
23. Lan, C.W., Li, B., Zhou, J.: Simultaneously concentrated electric and thermal fields using fan-shaped structure. Opt. Express **23**, 24475 (2015)
24. Yang, T.Z., Bai, X., Gao, D.L., Wu, L.Z., Li, B.W., Thong, J.T.L., Qiu, C.W.: Invisible sensors: Simultaneous sensing and camouflaging in multiphysical fields. Adv. Mater. **27**, 7752–7758 (2015)
25. Shen, X.Y., Li, Y., Jiang, C.R., Ni, Y.S., Huang, J.P.: Thermal cloak-concentrator. Appl. Phys. Lett. **109**, 031907 (2016)
26. Yang, W.Z., Liu, Q.C., Gao, Z.Z., Yue, Z.F., Xu, B.X.: Theoretical search for heterogeneously architected 2D structures. Proc. Natl. Acad. Sci. USA **115**, E7245 (2018)
27. Mitsumoto, K., Yoshino, H.: Orientational ordering of closely packed Janus particles. Soft Matter **14**, 3919–3928 (2018)
28. Fei, W.J., Driscoll, M.M., Chaikin, P.M., Bishop, K.J.M.: Magneto-capillary dynamics of amphiphilic Janus particles at curved liquid interfaces. Soft Matter **14**, 4661–4665 (2018)
29. Debnath, T., Li, Y.Y., Ghosh, P.K., Marchesoni, F.: Hydrodynamic interaction of trapped active Janus particles in two dimensions. Phys. Rev. E **97**, 042602 (2018)
30. Yang, S., Xu, L.J., Wang, R.Z., Huang, J.P.: Full control of heat transfer in single-particle structural materials. Appl. Phys. Lett. **111**, 121908 (2017)
31. Xu, L.J., Jiang, C.R., Shang, J., Wang, R.Z., Huang, J.P.: Periodic composites: quasi-uniform heat conduction, Janus thermal illusion, and illusion thermal diodes. Eur. Phys. J. B **90**, 221 (2017)
32. Xu, L.J., Yang, S., Huang, J.P.: Thermal theory for heterogeneously architected structure: fundamentals and application. Phys. Rev. E **98**, 052128 (2018)

Chapter 16
Theory for Anisotropic Core and Anisotropic Shell: Thermal Transparency, Concentrator and Cloak

Abstract In the existing literatures of thermal metamaterials or metadevices, many properties or functions are designed via coordinate transformation theory (transformation thermotics), including thermal concentrating and cloaking. But other properties or functions, say, thermal transparency, are designed by using theories differing from the transformation thermotics. Here we put forward an effective medium theory in thermotics by considering anisotropic layered/graded structures, and we reveal that the theory can unify transparency, concentrating, and cloaking into the same theoretical framework. Furthermore, the theory not only gives the criterion of transparency, concentrating, and cloaking, but also helps to predict a type of ellipses-embedded structures which can achieve transparency, concentrating, and cloaking, respectively. The prediction is confirmed by our finite-element simulations and/or experiments. This chapter provides a different theory to understand and design thermal metamaterials or metadevices, which might be extended into other disciplines, such as optics/electromagnetics and acoustics.

Keywords Effective medium theory · Thermal transparency · Thermal concentrator · Thermal cloak

16.1 Opening Remarks

Since 2008, thermal metamaterials or metadevices [1–22] have been intensively studied, in order to achieve invisibility [1–13], illusion [15–21] and other inconceivable thermal properties or functions, such as concentrators [4, 5, 11, 12], macroscopic diodes [10] and energy-free thermostats [22].

On the one hand, most of the devices are designed based on the theory of coordinate transformation [1–6, 10–12, 18–20], which originates from the pioneering work on electromagnetic waves in 2006 [23]. For example, this theory helps to predict and realize the effect of thermal cloaking (which helps to let the heat flow around an object as if the object does not exist) [1–12] and concentrating (which corresponds to the concentration of the heat into a specific region) [4, 5, 11, 12].

© Springer Nature Singapore Pte Ltd. 2020
J.-P. Huang, *Theoretical Thermotics*,
https://doi.org/10.1007/978-981-15-2301-4_16

On the other hand, some thermal metamaterials with other properties or functions are designed by using theories beyond transformation. Typically, thermal transparency (which means that heat flows across a region without disturbing outside thermal signatures) [13, 14] was proposed and designed by introducing the concept of neutral inclusion with computable effective thermal conductivities.

Here, we raise a question: does there exist a thermal theory which is capable of unifying thermal transparency, concentrating and cloaking into the same theoretical framework? If so, one would be able to design transparency, concentrating and cloaking from a different aspect, which may be convenient. Nevertheless, since thermal metamaterials with different properties or functions are composed of different structures and materials, it seems difficult to unify the theories. As an initial work, here we attempt to overcome the difficulty through developing an effective medium theory in thermotics, which considers anisotropic layered/graded structures. Our simulations show that the theory can not only unify thermal transparency, concentrating and cloaking into the same theoretical framework, but also help to design practical devices on the basis of an ellipses-embedded structure according to the resulting theoretical criterion. The desired effects are confirmed by simulation and/or experiment.

16.2 Theoretical Analysis of Two-Dimensional Circular Structures Constructed by Anisotropic Materials

16.2.1 Exact Solution for a Multi-layered Structure

First of all, we consider a bilayer structure which is composed of two anisotropic materials (see Fig. 16.1a). The thermal conductivities of them are second-order diagonal tensors which have radial element κ_{rr} (radial thermal conductivity) and tangential element $\kappa_{\theta\theta}$ (tangential thermal conductivity). Considering the structure presented in a uniform thermal gradient field without heat sources, the conduction equation satisfies the Laplace equation

$$\nabla \cdot (-\kappa \cdot \nabla T) = 0. \tag{16.1}$$

We conduct variable separation and derive the expression in polar coordinates since the thermal conductivity tensors have no items, $\kappa_{r\theta}$ and $\kappa_{\theta r}$. Then, we obtain

$$\frac{1}{r}\frac{\partial}{\partial r}\left(r\kappa_{rr}\frac{\partial T}{\partial r}\right) + \frac{1}{r}\frac{\partial}{\partial \theta}\left(\frac{\kappa_{\theta\theta}}{r}\frac{\partial T}{\partial \theta}\right) = 0. \tag{16.2}$$

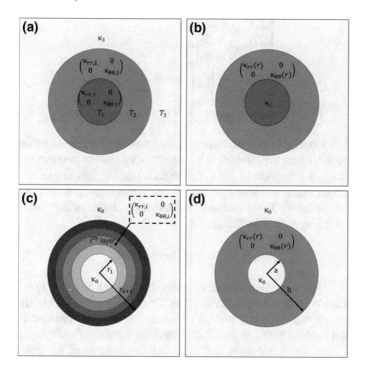

Fig. 16.1 Schematic diagram of **a** bilayer and **b** graded structures, **c** multi-layered and **d** graded rings. **a** For the bilayer structure, the circular core and annular shell are composed of anisotropic materials, with radial ($\kappa_{rr,1}$ and $\kappa_{rr,2}$) and tangential ($\kappa_{\theta\theta,1}$ and $\kappa_{\theta\theta,2}$) thermal conductivities, respectively. **b** The graded structure is composed of a circular core with uniform thermal conductivity κ_c and an annular shell with radial ($\kappa_{rr}(r)$) and tangential ($\kappa_{\theta\theta}(r)$) thermal conductivities, respectively. The multi-layered ring in **c** is composed of n layers of concentric rings. The i-th ring with inner radius r_i is composed of material with radial ($\kappa_{rr,i}$) and tangential ($\kappa_{\theta\theta,i}$) thermal conductivities, respectively. The graded ring with internal radius a and external radius b is composed of graded material with radial ($\kappa_{rr}(r)$) and tangential ($\kappa_{\theta\theta}(r)$) thermal conductivities, respectively. Adapted from Ref. [24]

The general solution of the above equation is

$$T = A_0 + B_0 \ln r + \sum_{m=1}^{\infty} [A_m \cos(m\theta) + B_m \sin(m\theta)] r^m \sqrt{\frac{\kappa_{\theta\theta}}{\kappa_{rr}}}$$

$$+ \sum_{n=1}^{\infty} [C_n \cos(n\theta) + D_n \sin(n\theta)] r^{-n} \sqrt{\frac{\kappa_{\theta\theta}}{\kappa_{rr}}}. \tag{16.3}$$

When the structure shown in Fig. 16.1a is presented in a background with thermal conductivity κ_3 and uniform thermal gradient field ∇T (along x direction), we can write the associated boundary conditions as

$$\begin{cases}
T_1\big|_{r\to 0} \text{ is finite,} \\
T_1\big|_{r=r1} = T_2\big|_{r=r1}, \\
T_2\big|_{r=r2} = T_3\big|_{r=r2}, \\
-\kappa_{rr,1}\dfrac{\partial T_1}{\partial r}\Big|_{r=r1} = -\kappa_{rr,2}\dfrac{\partial T_2}{\partial r}\Big|_{r=r1}, \\
-\kappa_{rr,2}\dfrac{\partial T_2}{\partial r}\Big|_{r=r2} = -\kappa_3\dfrac{\partial T_3}{\partial r}\Big|_{r=r2}, \\
T_3\big|_{r\to\infty} = -\nabla T r \cos\theta.
\end{cases} \qquad (16.4)$$

T_1, T_2 and T_3 represent the temperature of the core, shell and background, respectively. Since the device is presented in an external temperature gradient along x axis, the last boundary condition in Eq. (16.4) means that the thermal gradient field is uniform along x axis ($r\cos\theta = x$) at infinity (the device can't affect the temperature gradient at the infinity). Then, we can write down the solution of Eq. (16.2) by taking into account the boundary conditions,

$$T_3 = \left(A_1 r + B_1 r^{-1}\right)\cos\theta. \qquad (16.5)$$

Equation (16.5) partially keeps the first-order terms ($m = 1$ and $n = 1$) in Eq. (16.3) (general solution) to match the boundary conditions in Eq. (16.4). When the effective thermal conductivity of the structure (κ_e) is equal to that of the background (κ_3), there is no thermal contrast between the structure and background. Hence, B_1 is expected to be zero. Thus, we obtain the effective thermal conductivity of the structure,

$$\kappa_e = c_2\kappa_{rr,2}\frac{(1 + p^{c_2})c_1\kappa_{rr,1} + (1 - p^{c_2})c_2\kappa_{rr,2}}{(1 - p^{c_2})c_1\kappa_{rr,1} + (1 + p^{c_2})c_2\kappa_{rr,2}}, \qquad (16.6)$$

where $p = r_1^2/r_2^2$, $c_1 = \sqrt{\kappa_{\theta\theta,1}/\kappa_{rr,1}}$, and $c_2 = \sqrt{\kappa_{\theta\theta,2}/\kappa_{rr,2}}$. When $c_1 = c_2 = 1$ (isotropic material), Eq. (16.6) is reduced to the known Maxwell-Garnett theory.

So far, we have deduced the effective thermal conductivity of the bilayer circular structure which is composed of two anisotropic materials. For the multi-layered circular structure, we can firstly calculate the effective thermal conductivity (we define it as κ_{e1}) of the innermost two layers using the above conclusion. Then, by regarding the innermost two layers as one layer which possesses the uniform thermal conductivity of κ_{e1}, we can calculate the effective thermal conductivity of the innermost three layers. This procedure allows us to derive the effective thermal conductivity of the whole multi-layered structure by continuous iteration.

16.2.2   Exact Solution for a Graded Structure

Based on the above deduction, our theory can also be extended into calculating the effective thermal conductivity of a graded structure. The graded structure is composed of continuous medium whose anisotropic thermal conductivity varies along the radius. Considering a simple structure; see Fig. 16.1b, which is composed of a homogeneous circular core (with thermal conductivity of κ_c) and an anisotropic annular shell (with radial and tangential thermal conductivity of $\kappa_{rr}(r)$ and $\kappa_{\theta\theta}(r)$, respectively). By solving the Laplace equation, we obtain the effective thermal conductivity (κ_e) of the graded structure

$$\kappa_e = c\kappa_{rr}(r) \frac{(1 + p^c)\,\kappa_c + (1 - p^c)\,c\kappa_{rr}(r)}{(1 - p^c)\,\kappa_c + (1 + p^c)\,c\kappa_{rr}(r)}, \tag{16.7}$$

where p is the area fraction of the core and c is $\sqrt{\kappa_{\theta\theta}(r)/\kappa_{rr}(r)}$. For convenience, we rewrite Eq. (16.7) as

$$\frac{\kappa_e - \kappa_{rr}(r)c}{\kappa_e + \kappa_{rr}(r)c} = p^c \frac{\kappa_c - \kappa_{rr}(r)c}{\kappa_c + \kappa_{rr}(r)c}. \tag{16.8}$$

For a shell with infinitesimal thickness of dr encircling the graded structure, the effective thermal conductivity changes from $\kappa_e(r)$ to $\kappa_e(r + dr)$. In this case, Eq. (16.8) helps to obtain

$$\frac{\kappa_e(r + dr) - \kappa_{rr}(r)c}{\kappa_e(r + dr) + \kappa_{rr}(r)c} = \left[\frac{r^2}{(r + dr)^2}\right]^c \frac{\kappa_e(r) - \kappa_{rr}(r)c}{\kappa_e(r) + \kappa_{rr}(r)c}. \tag{16.9}$$

As a result, we obtain a differential equation

$$\frac{d\kappa_e(r)}{dr} = \frac{[c\kappa_{rr}(r)]^2 - \kappa_e(r)^2}{r\kappa_{rr}(r)}. \tag{16.10}$$

Given the gradation profiles $[\kappa_{rr}(r)$ and $\kappa_{\theta\theta}(r)]$ and the boundary condition (when the radius is close to zero), the effective thermal conductivity of the whole graded structure, $\kappa_e(r)$, can be calculated according to Eq. (16.10).

16.2.3   Criterion for Transparency, Concentrating and Cloaking

So far, we have calculated the effective thermal conductivities of layered and graded structures which are composed of anisotropic materials. Based on these calculations, we are now in a position to provide the criterion of thermal transparency, concentrating and cloaking, and then these devices can be designed by using anisotropic materials accordingly. Let us consider two rings which are composed of multi-layered (Fig. 16.1c) and graded (Fig. 16.1d) anisotropic materials embedded in backgrounds with thermal conductivity κ_0. The following criterions hold true when $\kappa_{rr,i}\kappa_{\theta\theta,i} = \kappa_0^2$ (for multi-layered materials) and $\kappa_{rr}(r)\kappa_{\theta\theta}(r) = \kappa_0^2$ (for graded materials) are satisfied. Namely,

(1) When $\kappa_{rr}^{eff} = \kappa_{\theta\theta}^{eff}$, the structures shown in Fig. 16.1c, d are serving as thermal transparency devices.

(2) When $\kappa_{rr}^{eff} > \kappa_{\theta\theta}^{eff}$, the structures shown in Fig. 16.1c, d are serving as thermal concentrators.

(3) When $\kappa_{rr}^{eff} < \kappa_{\theta\theta}^{eff}$, the structures shown in Fig. 16.1c, d are serving as thermal cloaks. Especially, they tend to be perfect cloaks when $\kappa_{rr}^{eff} \to 0$.

κ_{rr}^{eff} and $\kappa_{\theta\theta}^{eff}$ are the effective radial and tangential thermal conductivities respectively. For the multi-layered structure described in Fig. 16.1c, we have

$$\kappa_{rr}^{eff} = \ln \frac{r_{n+1}}{r_1} \left(\sum_{i=1}^{n} \frac{1}{\kappa_{rr,i}} \ln \frac{r_{i+1}}{r_i} \right)^{-1},$$

$$\kappa_{\theta\theta}^{eff} = \left(\ln \frac{r_{n+1}}{r_1} \right)^{-1} \sum_{i=1}^{n} \kappa_{\theta\theta,i} \ln \frac{r_{i+1}}{r_i}. \tag{16.11}$$

For the graded structure described in Fig. 16.1d, we obtain

$$\kappa_{rr}^{eff} = \ln \frac{b}{a} \left(\int_{a}^{b} \frac{dr}{\kappa_{rr}(r)r} \right)^{-1},$$

$$\kappa_{\theta\theta}^{eff} = \left(\ln \frac{b}{a} \right)^{-1} \int_{a}^{b} \frac{\kappa_{\theta\theta}(r)dr}{r}. \tag{16.12}$$

16.3   Design of Thermal Transparency Devices, Concentrators and Cloaks via a Finite-Element Method

According to the above theory, we firstly design two thermal transparency devices based on two-dimensional simulations by using the solid heat transfer module of commercial software COMSOL (https://www.comsol.com). The physics-controlled mesh is adjusted to be extremely fine in each simulation model. The left and right sides are fixed at constant temperatures playing the roles of heat and cold sources. The boundary conditions of upper and bottom sides are heat insulation. Besides, in our simulations, the air convection is not considered. Figure 16.2a, b shows the simulation results of an anisotropic (Fig. 16.2a) and a graded-anisotropic (Fig. 16.2b) annular devices with internal diameter 2 cm and external diameter 4 cm embedded in square hosts with side length 9 cm. The basic parameters are described in the figure caption. To maintain uniform densities of heat flux, the left sides of the hosts hold linear hot sources with temperature 323 K, and the right sides are linear cold sources with temperature 273 K. The color surfaces display the distribution of the temperature, and the white lines represent the isotherms. The isotherms inside and

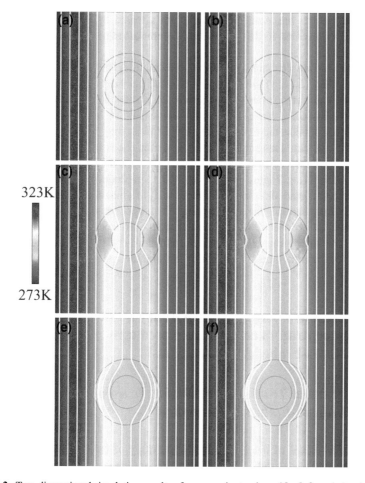

Fig. 16.2 Two-dimensional simulation results of **a, c, e** anisotropic and **b, d, f** graded-anisotropic thermal transparency devices, concentrators and cloaks presented in uniform backgrounds with thermal conductivity 30 Wm^{-1}K^{-1}. **a** The anisotropic transparency device is composed of two layers of materials with the radial (36 and 24.293 Wm^{-1}K^{-1}) and tangential (25 and 37.047 Wm^{-1}K^{-1}) thermal conductivities, respectively. **b** The graded-anisotropic transparency device is composed of graded material with the radial ($\frac{900}{(29\ln 2)r-1}$ Wm^{-1}K^{-1}) and tangential ($(29\ln 2)r - 1$ Wm^{-1}K^{-1}) thermal conductivities, respectively. **c** The anisotropic concentrator is composed of material with the radial (100 Wm^{-1}K^{-1}) and tangential (9 Wm^{-1}K^{-1}) thermal conductivities, respectively. **d** For the graded anisotropic concentrator, the ring is composed of graded material with the radial ($100r - 30$ Wm^{-1}K^{-1}) and tangential ($\frac{900}{100r-30}$ Wm^{-1}K^{-1}) thermal conductivities, respectively. The parameters of the two cloaks (**e, f**) are same as those used in the two concentrators (**c, d**) respectively, except the interchanging parameters of tangential and radial thermal conductivities. Adapted from Ref. [24]

outside the devices are straight (which is not affected by the devices) as if there is no devices presented in the backgrounds, demonstrating the behavior of thermal transparency.

Then, we further design two thermal concentrators (Fig. 16.2c, d) and two thermal cloaks (Fig. 16.2e, f) which have the same geometrical parameters with that discussed in transparency devices. Figure 16.2c, d show the simulation results of an anisotropic (Fig. 16.2c) and a graded-anisotropic (Fig. 16.2d) thermal concentrators. The isotherms outside the devices are straight reflecting the zero thermal contrast between the concentrators and the hosts. However, the isotherms inside the devices are bent and compressed to the center of the devices, exhibiting the phenomenon of thermal concentrating. Figure 16.2e, f shows the simulation results of an anisotropic (Fig. 16.2e) and a graded-anisotropic (Fig. 16.2f) thermal cloaks. In order to prevent the heat flux from entering the cloak regions, the tangential thermal conductivities are set to be much larger than the radial ones. The straight isotherms outside the devices reflect that the outer temperature gradients can hardly affect the cloak regions as the heat flows around the inner cloak regions through the anisotropic mediums.

16.4   Design of Thermal Transparency Devices, Concentrators and Cloaks Based on Ellipses-Embedded Structures

16.4.1   Thermal Transparency Device Based on an Ellipses-Embedded Structure

By taking advantage of the geometrical anisotropy of the air ellipses, we further design a thermal transparency device based on an ellipses-embedded structure; see Fig. 16.3b, c, d, verifying the above theory in both experiment and simulation. The thermal transparency device is composed of four concentric rings with multiple air ellipses embedded. The basic parameters of the structure are depicted in the figure caption. According to Ref. [21], we can achieve the effective thermal conductivity of a two-dimensional ellipse-embedded square host along the i-th axis of the ellipse,

$$\kappa_i = \kappa_n \frac{[p + (1 - p) L_i] \kappa_m + (1 - p) (1 - L_i) \kappa_n}{(1 - p) L_i \kappa_m + [p + (1 - p) (1 - L_i)] \kappa_n}. \tag{16.13}$$

Here, we consider a binary composite where an air ellipse of thermal conductivity κ_m and area fraction p is embedded in a host medium of κ_n, in the presence of a uniform external thermal gradient field. To be mentioned, L_i ($i = a, b$) is the shape factor of the two-dimensional ellipse, while a and b are the semi-axes of the air ellipses. We have

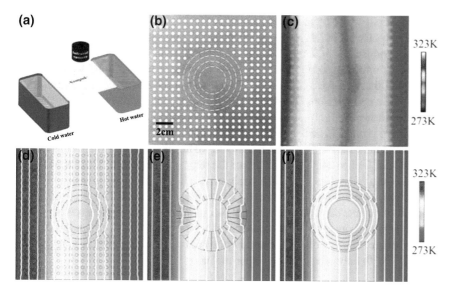

Fig. 16.3 Experimental **a** setup and **b** sample of thermal transparency device. The transparency device is composed of 100 air ellipses, each with major (minor) semi-axis 0.33 cm (0.04 cm), which are drilled in brass of 109 $Wm^{-1}K^{-1}$. The distance between nearby ellipses equals to 0.83 cm. The host background is occupied by brass drilled with 312 air circles with radius 0.23 cm, providing thermal conductivity 80 $Wm^{-1}K^{-1}$. Two-dimensional experimental and simulation results of ellipses-embedded structures: **c**, **d** thermal transparency device, **e** concentrator, and **f** cloak. **c** Shows the experimental result of a transparency device (**d**) is the simulation result corresponding to (**c**). **e** Shows the simulation result of the concentrator in the presence of a uniform background with the thermal conductivity of 120 $Wm^{-1}K^{-1}$. The concentrator is composed of five layers of air ellipses, each with major (minor) semi-axis 0.3 cm (0.03 cm), which are embedded in the background 270 $Wm^{-1}K^{-1}$. The width of the concentrator equals 3 cm. **f** Shows the simulation result of the thermal cloak. The cloak is composed of five layers of air ellipses embedded in the background of 380 $Wm^{-1}K^{-1}$; their major semi-axes are 0.3 cm, 0.355 cm, 0.415 cm, 0.48 cm and 0.54 cm respectively, and their minor semi-axes are all 0.05 cm. Other parameters are same as those used in (**c**). Adapted from Ref. [24]

$$L_i = \frac{ab}{2} \int_0^\infty \frac{ds}{\left(i^2 + s\right)\sqrt{\left(a^2 + s\right)\left(b^2 + s\right)}}. \tag{16.14}$$

Then, we extend the above results into a two-dimensional ring which contains many air ellipses. It is easy to verify that as long as the distances between nearby ellipses are equal to the ring's width and one of the axes of the air ellipses is set along the ring's radius, the radial and tangential thermal conductivities of the ring can be approximately calculated by using Eq. (16.13). Thus, for the thermal transparency device depicted in Fig. 16.3b, c, d, the calculated results of the radial and tangential thermal conductivities of the four rings are 68 $Wm^{-1}K^{-1}$ and 101 $Wm^{-1}K^{-1}$, respectively. Hence, the effective thermal conductivity of the whole device can be obtained by utilizing the above theory, which is equal to 84 $Wm^{-1}K^{-1}$.

In addition, we also conduct an experiment to verify the above calculation. The sample (Fig. 16.3b) is fabricated by chemical etching of 0.03 cm brass board with thermal conductivity 109 $Wm^{-1}K^{-1}$, and 0.1 mm-thick polydimethylsiloxane film is covered on the sample in order to eliminate the infrared refection. A plastic foam board with the same size of sample is stuck on the back of the sample in order to weaken the air convection. Two water tanks, which are respectively filled with hot and ice water (shown in Fig. 16.3a), serve as heat and cold sources, respectively. The room temperature is tuned to be the middle temperature between the heat and cold sources, which can minimize the air convection. Then we use FLIR E60 infrared camera with resolution of 240×240 pixels to detect the temperature profile. The experimental result is displayed in Fig. 16.3c. The straight isotherms reflect that there is no thermal contrast between the device and host background; that is to say, the effective thermal conductivity of the device is the same as that of background. In the experiment, the thermal conductivity of the background is set to be 80 $Wm^{-1}K^{-1}$, which is nearly the same with the above calculated result. Figure 16.3d is the simulation result, which is consistent with the corresponding experimental result.

16.4.2   Thermal Concentrator and Cloak Based on Ellipses-Embedded Structures

By using the simulation method, we further design a thermal concentrator and a cloak with the radius of 5.5 cm based on the ellipses-embedded structures. To be different from the above equidistant distribution of the air ellipses, the air ellipses are equal-angle-distributed in the annular structures. The detail parameters are described in the figure caption. Figure 16.3e, f show the two-dimensional simulation results of the thermal concentrator and cloak. The white isotherms clearly show that the thermal concentrating (Fig. 16.3e) and cloaking effect (Fig. 16.3f) are achieved. In principle, thermal concentrator and cloak can be experimentally fabricated on the basis of the two designs. However, the thermal contact resistance of the welded junctions (which connect different materials) must be carefully eliminated, in order to keep the performance of the samples.

16.5   Conclusion

To sum up, we have proposed an effective medium theory in thermotics, which allows us to unify transparency, concentrating and cloaking into the same theoretical framework. The resulting theoretical criterion has helped us to design transparency, concentrating and cloaking, and we have confirmed the three functions via finite-element simulations. Furthermore, with the aid of the theory, we have introduced an ellipses-embedded structure for transparency, concentrating and cloaking; the desired effects have been verified in simulations and/or experiment. Our theory and the corresponding ellipses-embedded structure may be applied to achieve other thermal metamaterials like rotators, which are practically and commercially available for potential applications. In addition, our theory, together with the criterion, might be extended into other disciplines, such as optics/electromagnetics and acoustics.

16.6   Exercises and Solutions

Exercises

1. Consider a two-dimensional core-shell structure. The core is with radius r_c and scalar thermal conductivity κ_c, and the shell is with radius r_s and nonuniform tensorial thermal conductivity $\kappa_s = \text{diag}\,[\kappa_{rr}\,(r)\,,\ \kappa_{\theta\theta}\,(r)]$. Please derive the differential equation to calculate the effective thermal conductivity of the core-shell structure κ_e. (Note that if the shell is uniform, the effective thermal conductivity of the core-shell can be calculated as $\kappa_e = m\kappa_{rr}\frac{\kappa_c+m\kappa_{rr}+(\kappa_c-m\kappa_{rr})p^m}{\kappa_c+m\kappa_{rr}-(\kappa_c-m\kappa_{rr})p^m}$, where $m = \sqrt{\kappa_{\theta\theta}/\kappa_{rr}}$ and $p = (r_c/r_s)^2$.)

Solutions

1. **Solution**: For convenience, we rewrite the given equation as $\frac{\kappa_e-m\kappa_{rr}}{\kappa_e+m\kappa_{rr}} = p^m \frac{\kappa_c-m\kappa_{rr}}{\kappa_c+m\kappa_{rr}}$. Considering a shell with infinitesimal thickness of dr encircling the graded structure, the effective thermal conductivity changes from $\kappa_e(r)$ to $\kappa_e(r+dr)$. In this case, we can obtain $\frac{\kappa_e(r+dr)-m\kappa_{rr}(r)}{\kappa_e(r+dr)+m\kappa_{rr}(r)} = \left[\frac{r^2}{(r+dr)^2}\right]^m \frac{\kappa_e(r)-m\kappa_{rr}(r)}{\kappa_e(r)+m\kappa_{rr}(r)}$. Thus, we obtain a differential equation $\frac{d\kappa_e(r)}{dr} = \frac{[m\kappa_{rr}(r)]^2-\kappa_e(r)^2}{r\kappa_{rr}(r)}$.

References

1. Fan, C.Z., Gao, Y., Huang, J.P.: Shaped graded materials with an apparent negative thermal conductivity. Appl. Phys. Lett. **92**, 251907 (2008)
2. Chen, T.Y., Weng, C.N., Chen, J.S.: Cloak for curvilinearly anisotropic media in conduction. Appl. Phys. Lett. **93**, 114103 (2008)
3. Li, J.Y., Gao, Y., Huang, J.P.: A bifunctional cloak using transformation media. J. Appl. Phys. **108**, 074504 (2010)
4. Guenneau, S., Amra, C., Veynante, D.: Transformation thermodynamics: cloaking and concentrating heat flux. Opt. Express **20**, 8207–8218 (2012)
5. Narayana, S., Sato, Y.: Heat flux manipulation with engineered thermal materials. Phys. Rev. Lett. **108**, 214303 (2012)
6. Schittny, R., Kadic, M., Guenneau, S., Wegener, M.: Experiments on transformation thermodynamics: molding the flow of heat. Phys. Rev. Lett. **110**, 195901 (2013)
7. Xu, H.Y., Shi, X.H., Gao, F., Sun, H.D., Zhang, B.L.: Ultrathin three-dimensional thermal cloak. Phys. Rev. Lett. **112**, 054301 (2014)
8. Han, T.C., Bai, X., Gao, D.L., Thong, J.T.L., Li, B.W., Qiu, C.-W.: Experimental demonstration of a bilayer thermal cloak. Phys. Rev. Lett. **112**, 054302 (2014)

9. Ma, Y.G., Liu, Y.C., Raza, M., Wang, Y.D., He, S.L.: Experimental demonstration of a mul-
    tiphysics cloak: manipulating heat flux and electric current simultaneously. Phys. Rev. Lett.
    113, 205501 (2014)
10. Li, Y., Shen, X.Y., Wu, Z.H., Huang, J.Y., Chen, Y.X., Ni, Y.S., Huang, J.P.: Temperature-
    dependent transformation thermotics: from switchable thermal cloaks to macroscopic thermal
    diodes. Phys. Rev. Lett. **115**, 195503 (2015)
11. Chen, T.Y., Weng, C.N., Tsai, Y.L.: Materials with constant anisotropic conductivity as a
    thermal cloak or concentrator. J. Appl. Phys. **117**, 054904 (2015)
12. Shen, X.Y., Li, Y., Jiang, C.R., Ni, Y.S., Huang, J.P.: Thermal cloak-concentrator. Appl. Phys.
    Lett. **109**, 031907 (2016)
13. He, X., Wu, L.Z.: Thermal transparency with the concept of neutral inclusion. Phys. Rev. E **88**,
    033201 (2013)
14. Zeng, L.W., Song, R.X.: Experimental observation of heat transparency. Appl. Phys. Lett. **104**,
    201905 (2014)
15. Han, T.C., Bai, X., Thong, J.T.L., Li, B.W., Qiu, C.-W.: Full control and manipulation of heat
    signatures: cloaking, camouflage and thermal metamaterials. Adv. Mat. **26**, 1731–1734 (2014)
16. Yang, T.Z., Su, Y.S., Xu, W.K., Yang, X.D.: Transient thermal camouflage and heat signature
    control. Appl. Phys. Lett. **109**, 121905 (2016)
17. Yang, T.Z., Bai, X., Gao, D.L., Wu, L.Z., Li, B.W., Thong, J.T.L., Qiu, C.W.: Invisible sensors:
    simultaneous sensing and camouflaging in multiphysical fields. Adv. Mater. **27**, 7752–7758
    (2015)
18. Chen, X.Y., Shen, X.Y., Huang, J.P.: Engineering the accurate distortion of an object's
    temperature-distribution signature. Eur. Phys. J. Appl. Phys. **70**, 20901 (2015)
19. Zhu, N.Q., Shen, X.Y., Huang, J.P.: Converting the patterns of local heat flux via thermal
    illusion device. AIP Adv. **5**, 053401 (2015)
20. Hou, Q.W., Zhao, X.P., Meng, T., Liu, C.L.: Illusion thermal device based on material with con-
    stant anisotropic thermal conductivity for location camouflage. Appl. Phys. Lett. **109**, 103506
    (2016)
21. Yang, S., Xu, L.J., Wang, R.Z., Huang, J.P.: Full control of heat transfer in single-particle
    structural materials. Appl. Phys. Lett. **111**, 121908 (2017)
22. Shen, X.Y., Li, Y., Jiang, C.R., Huang, J.P.: Temperature trapping: energy-free maintenance of
    constant temperatures as ambient temperature gradients change. Phys. Rev. Lett. **117**, 055501
    (2016)
23. Pendry, J.B., Schurig, D., Smith, D.R.: Controlling electromagnetic fields. Science **312**, 1780–
    1782 (2006)
24. Wang, R.Z., Xu, L.J., Ji, Q., Huang, J.P.: A thermal theory for unifying and designing trans-
    parency, concentrating and cloaking. J. Appl. Phys. **123**, 115117 (2018)

Chapter 17
Theory for Periodic Structure: Thermal Transparency

Abstract Almost all thermal metamaterials are essentially achieved by tailoring asymmetric interaction between matrices and embedded particles. However, the asymmetric interaction results in the noncommutativity of matrices and particles, which may reduce the flexibility for heat management. To solve this problem, this chapter describes a different mechanism by tailoring symmetric interaction between particles arranged in periodic lattices, thus being called periodic interparticle interaction. For practical application, the representative thermal transparency is introduced, which, however, is realized by tailoring periodic interparticle interaction. Theoretical analysis, finite-element simulation, and laboratory experiment all validate the proposed mechanism. Moreover, the Maxwell–Garnett theory and the Bruggeman theory are re-visited from their scope and relation with theoretical analysis and finite-element analysis. Compared with the existing thermal transparency, the present scheme looks more feasible to handle many-particle systems. This chapter opens a gate to exploring periodic interparticle interaction, and further work can be expected: (I) exploring periodic interparticle interaction with different lattice types and relative positions for particle arrangement; (II) applying periodic interparticle interaction to achieve other functions, such as thermal camouflage.

Keywords Periodic structure · Symmetric interaction · Thermal transparency · Maxwell–Garnett theory · Bruggeman theory

17.1 Opening Remarks

Thermal metamaterials have attracted wide research interest since 2008 [1–3]. The related mechanism is mainly to tailor asymmetric interaction between matrices and embedded particles. For example, thermal transparency in Refs. [4–7] considers the asymmetric interaction between shell and inside core; thermal invisibility in Refs. [8, 9] tailors the asymmetric interaction between matrix and inside particles. The two phenomena are featured by the same thermal conductivities between the device and background. For core-shell structure, background is the region outside the shell; for matrix plus inside particles, background is the region outside the matrix. Therefore,

© Springer Nature Singapore Pte Ltd. 2020
J.-P. Huang, *Theoretical Thermotics*,
https://doi.org/10.1007/978-981-15-2301-4_17

background can be generally regarded as the region excluding the device (say, the designed structure). Note that the background may also possess uniform microstructures. Other thermal metamaterials for heat management, such as thermal cloak [1, 2, 10–14], thermal concentrator [10, 15, 16], and thermal camouflage [17–23], are all in the framework of asymmetric interaction. Although asymmetric interaction has been well manipulated, the noncommutativity of matrices and inside particles may restrict the flexibility for heat management.

To solve this problem, here we reveal the mechanism of symmetric interaction between periodic particles, say periodic interparticle interaction. For practical application, we take the representative thermal transparency in Refs. [4–7] as an example, but realize the phenomenon beyond the asymmetric interaction between shell and inside core (an equivalent expression is the neutral inclusion which was raised by Ref. [4]). Instead, we carefully tailor periodic interparticle interaction to remove the influence of periodic particles [say, particles A and particles B in Fig. 17.1a] on the background, thus thermal transparency is also achieved. Here particles A and particles B are the designed structure, thus background is the region excluding particles A and particles B. Note that if the background is a pure material, it is just the matrix where particles A and particles B are embedded. If the background possesses microstructures, it can be regarded as the matrix plus microstructures. Then the scheme will be validated by theoretical analysis, finite-element simulation, and laboratory experiment.

Moreover, we find that the formula describing periodic interparticle interaction is mathematically the same as the well-known Bruggeman formula. Although the Bruggeman formula can indeed explain symmetric interaction, the understanding is completely different in our work, which will be discussed as well.

17.2 Theory for Periodic Interparticle Interaction

Let us start from considering the two-dimensional periodic composite material presented in Fig. 17.1a. Two types of particles are alternately embedded in the matrix with distance d_0; see Fig. 17.1b. Particle A is featured by material anisotropy and geometry isotropy (a red graded circle with radius r), whereas particle B is characterized by material isotropy and geometry anisotropy (a blue ellipse with semi-minor axis s and semi-major axis t); see Fig. 17.1c, d. We set the thermal conductivities of particle A, particle B, and matrix to be $\bar{\kappa}_a = \text{diag}\left(\kappa_{\rho\rho},\ \kappa_{\theta\theta}\right)$, κ_b, and κ_m, respectively. $\bar{\kappa}_a$ is written in the cylindrical coordinates $(\rho,\ \theta)$ whose origin locates in the center of particle A. In what follows, we focus on the horizontal properties of the periodic composite material except for additional statements. In the presence of a thermal field E_0 (defined as negative temperature gradient) along horizontal direction, the effective thermal conductivity of the periodic composite material κ_e can be calculated by Fourier's law,

$$\kappa_e = \frac{\langle J \rangle}{\langle E \rangle}, \tag{17.1}$$

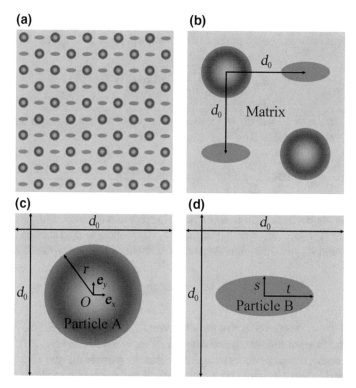

Fig. 17.1 Schematic diagrams of **a** periodic composite material and **b–d** basic structures. Adapted from Ref. [29]

where J and E represent the heat flux and thermal field in the periodic composite material, respectively. In Eq. (17.1), $\langle\cdots\rangle$ denotes the area average over the whole region. Considering the local heat fluxes (and local thermal fields) in particles A J_a (E_a), particles B J_b (E_b), and matrix J_m (E_m), Eq. (17.1) can be rewritten as

$$\kappa_e = \frac{p_a \langle J_a \rangle + p_b \langle J_b \rangle + p_m \langle J_m \rangle}{p_a \langle E_a \rangle + p_b \langle E_b \rangle + p_m \langle E_m \rangle}, \qquad (17.2)$$

where $p_a \left[= \pi r^2 / \left(2d_0^2\right)\right]$, $p_b \left[= \pi st / \left(2d_0^2\right)\right]$, and $p_m = 1 - p_a - p_b$ are the area fraction of particles A, particles B, and matrix, respectively.

Then we require to calculate the average heat fluxes $\langle J_a \rangle$, $\langle J_b \rangle$, and $\langle J_m \rangle$ in Eq. (17.2). Due to translation invariance, the calculations can be simplified to the two basic structures presented in Fig. 17.1c, d. Although the matrix in Fig. 17.1c, d is finite with size $d_0 \times d_0$, we can still make a reasonable and contributing approximation that the matrix is infinite, when the particles are small enough. The influence of infinite-matrix approximation will be analyzed after the theory for periodic interparticle interaction is established.

We firstly calculate the average heat flux in particle A; see Fig. 17.1c. For infinite-matrix approximation, the temperature distribution in particle A, T_a, can be expressed as [24]

$$T_a = \frac{-2r^{1-v} E_0 \kappa_m}{v \kappa_{\rho\rho} + \kappa_m} \rho^v \cos\theta + T_0, \tag{17.3}$$

where $v = \sqrt{\kappa_{\theta\theta}/\kappa_{\rho\rho}}$, and T_0 is a constant temperature. The average heat flux in particle A can be calculated by

$$\langle J_a \rangle = \frac{\iint_\Sigma \boldsymbol{e}_x \cdot (\bar{\kappa}_a \boldsymbol{E}_a) \, dS}{\iint_\Sigma dS} = \frac{\iint_\Sigma \left(\kappa_{xx} E_{ax} + \kappa_{xy} E_{ay} \right) dS}{\iint_\Sigma dS}, \tag{17.4}$$

where $\boldsymbol{e}_x$ is the horizontal unit vector, and Σ represents the closed integration area, i.e., particle A. $\kappa_{xx} = \kappa_{\rho\rho} \cos^2\theta + \kappa_{\theta\theta} \sin^2\theta$, $\kappa_{xy} = (\kappa_{\rho\rho} - \kappa_{\theta\theta}) \sin\theta \cos\theta$, $E_{ax} = -\partial T_a/\partial x$, and $E_{ay} = -\partial T_a/\partial y$. After calculating the integration, Eq. (17.4) can be reduce to

$$\langle J_a \rangle = \kappa_a \langle E_a \rangle = v\kappa_{\rho\rho} \frac{2E_0 \kappa_m}{v\kappa_{\rho\rho} + \kappa_m}, \tag{17.5}$$

where $\kappa_a \left(= v\kappa_{\rho\rho} \right)$ is the effective scalar thermal conductivity of particle A with the tensorial $\bar{\kappa}_a$. In what follows, we will use κ_a, if $\bar{\kappa}_a$ is not necessary.

We secondly calculate the average heat flux in particle B; see Fig. 17.1d. For infinite-matrix approximation, the temperature distribution in particle B, T_b, can be expressed as [25]

$$T_b = \frac{-E_0 \kappa_m}{\kappa_b L + \kappa_m (1 - L)} \rho \cos\theta + T_0, \tag{17.6}$$

where the horizontal shape factor L is given by

$$L = \frac{st}{2} \int_0^\infty \frac{du}{(t^2 + u) \sqrt{(s^2 + u)(t^2 + u)}}. \tag{17.7}$$

It is found that the thermal field in particle B is uniform, and hence the average heat flux in particle B can be directly expressed as

$$\langle J_b \rangle = \kappa_b \langle E_b \rangle = \kappa_b \frac{E_0 \kappa_m}{\kappa_b L + \kappa_m (1 - L)}. \tag{17.8}$$

We finally calculate the average heat flux in the matrix, and take the matrix in Fig. 17.1c as an example. For infinite-matrix approximation, the temperature distribution in the matrix T_m can be expressed as [24]

$$T_m = -E_0 \rho \cos\theta + \frac{r^2 E_0 (\kappa_a - \kappa_m)}{\kappa_a + \kappa_m} \rho^{-1} \cos\theta + T_0. \tag{17.9}$$

The average heat flux in the matrix is then determined by

$$\langle J_m \rangle = \kappa_m \langle E_m \rangle = \kappa_m E_0, \tag{17.10}$$

where $\langle E_m \rangle = E_0$ is because the effect of the term containing ρ^{-1} in Eq. (17.9) is local, which can be neglected due to infinite-matrix approximation. Equation (17.10) can also describe the average heat flux in the matrix as shown in Fig. 17.1d.

Considering Eqs. (17.5), (17.8), (17.10), we can rewrite Eq. (17.2) as

$$\kappa_e = \frac{p_a \kappa_a \langle E_a \rangle + p_b \kappa_b \langle E_b \rangle + p_m \kappa_m \langle E_m \rangle}{p_a \langle E_a \rangle + p_b \langle E_b \rangle + p_m \langle E_m \rangle} = \frac{p_a \eta_a \kappa_a + p_b \eta_b \kappa_b + p_m \kappa_m}{p_a \eta_a + p_b \eta_b + p_m}, \tag{17.11}$$

where $\eta_a = \langle E_a \rangle / \langle E_m \rangle = 2\kappa_m / (\kappa_a + \kappa_m)$ and $\eta_b = \langle E_b \rangle / \langle E_m \rangle = \kappa_m / [\kappa_b L + \kappa_m (1 - L)]$. The κ_e determined by Eq. (17.11) can describe the effective thermal conductivity of the whole system composed of particles A, particles B and the matrix. For physical understanding, Eq. (17.11) is, in a sense, the average of κ_a, κ_b, and κ_m with respect to $p_a \eta_a$, $p_b \eta_b$, and p_m. Note that the weight of κ_m (only with p_m) is different from those of κ_a and κ_b, which agrees with the comment that the interaction between the matrix and inside particles is asymmetric.

To exclude the thermal effect of the matrix, we further set the κ_e in Eq. (17.11) to be κ_m to consider the periodic interparticle interaction between particles A and particles B, and derive the only physical solution,

$$\kappa_m = \frac{-B + \sqrt{B^2 - 4AC}}{2A}, \tag{17.12}$$

where $A = p(1 - 2L) + 1$, $B = -\kappa_a [p(3 - 2L) - 1] + \kappa_b [p(1 + 2L) - 1]$, $C = \kappa_a \kappa_b [p(1 - 2L) - 1]$, $p = p_a / p_{a+b}$, and $p_{a+b} = p_a + p_b$. The κ_m determined by Eq. (17.12) can describe the periodic interparticle interaction between particles A and particles B, because the right-hand side of Eq. (17.12) is independent of the matrix property. In other words, particles A plus particles B have the same thermal property as the matrix according to Eq. (17.12). Therefore, thermal transparency can be realized by tailoring periodic interparticle interaction between particles A and particles B with the aid of Eq. (17.12). For completeness, we present the theory for three-dimensional thermal transparency in the Supplementary Proof.

17.3   Validating the Infinite-Matrix Approximation by Comparing with the Finite-Element Simulation

In the process to derive Eqs. (17.11), (17.12), we have adopted the infinite-matrix approximation, which assumes that particles are small enough. So we perform finite-element simulation to analyze the influence of area fractions on predicting the effective thermal conductivities. We consider the isotropic case with $\kappa_{\rho\rho} = \kappa_{\theta\theta}$ for

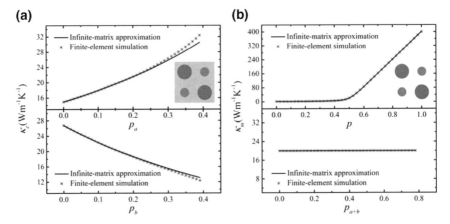

Fig. 17.2 Thermal conductivities of **a** periodic composite material κ_e and **b** periodic interparticle interaction κ_m, which are both given as a function of the area fraction of the particles. The solid lines in **a** [or **b**] are derived from Eq. (17.11) [or Eq. (17.12)], and the symbols are obtained with finite-element simulation. The size of the periodic composite material is $10 \times 10\,\text{cm}$, $d_0 = 1\,\text{cm}$, $\kappa_a = 400\,\text{Wm}^{-1}\text{K}^{-1}$, and $\kappa_b = 1\,\text{Wm}^{-1}\text{K}^{-1}$. Other parameters: **a** $\kappa_m = 20\,\text{Wm}^{-1}\text{K}^{-1}$, $p_b = 0.16$ for $\kappa_e - p_a$ curve, and $p_a = 0.16$ for $\kappa_e - p_b$ curve; **b** $p_{a+b} = 0.24$ for $\kappa_m - p$ curve, $p = 0.5$ for $\kappa_m - p_{a+b}$ curve, and the symbols are derived by putting the periodic particles into the matrix determined by Eq. (17.12) and calculating κ_m with $\langle J_{a+b} \rangle / \langle E_{a+b} \rangle$. Adapted from Ref. [29]

particle A and $s = t$ for particle B. κ_a and κ_b are set to be 400 and $1\,\text{Wm}^{-1}\text{K}^{-1}$, respectively. We compare the values of κ_e and κ_m derived from the infinite-matrix approximation [Eqs. (17.11), (17.12)] and finite-element simulation based on the software COMSOL Multiphysics (http://www.comsol.com/).

When discussing the influences of p_a and p_b on κ_e, we construct the periodic composite material with 5×5 array of the basic structure presented in Fig. 17.2a. κ_m is set to be $20\,\text{Wm}^{-1}\text{K}^{-1}$. The solid lines in Fig. 17.2a are derived from Eq. (17.11), which result from the infinite-matrix approximation. Then we put the periodic composite material into a thermal field, and derive the values of $\langle J \rangle$ and $\langle E \rangle$ with the finite-element simulation. We further calculate κ_e with $\langle J \rangle / \langle E \rangle$, and the results are presented by the symbols in Fig. 17.2a. The maximum value of p_a (and p_b) is $\pi/8$ (≈ 0.39), for a circle cannot fill the whole square. Clearly, the infinite-matrix approximation underestimates the effect of particle A (high thermal conductivity $400\,\text{Wm}^{-1}\text{K}^{-1}$) about 5.8% at $p_a = 0.39$, and overestimates the effect of particle B (low thermal conductivity $1\,\text{Wm}^{-1}\text{K}^{-1}$) about 6.2% at $p_b = 0.39$.

The results obtained from Fig. 17.2a illustrate that the infinite-matrix approximation is still contributing in spite of the large area fractions. The main reason is that the large area fractions (against infinite-matrix approximation) do change the theoretically predicted temperature distributions T_a, T_b, and T_m [Eqs. (17.3), (17.6), (17.9)], but we only focus on their average effects when calculating the average heat fluxes $\langle J_a \rangle$, $\langle J_b \rangle$, and $\langle J_m \rangle$ [Eqs. (17.5), (17.8), (17.10)]. As a result, the influence of large area fractions is reduced.

Then we discuss the influence of p and p_{a+b} on κ_m; see Fig. 17.2b. We construct the periodic particles with 5×5 array of the basic structure. The solid lines in Fig. 17.2b are derived from Eq. (17.12) (infinite-matrix approximation). Then we put these periodic particles into the matrix determined by Eq. (17.12) to analyze the practical κ_m. We derive the average heat flux $\langle J_{a+b} \rangle$ and average thermal field $\langle E_{a+b} \rangle$ in particles A and particles B with the finite-element simulation. We further calculate the κ_m with $\langle J_{a+b} \rangle / \langle E_{a+b} \rangle$, and the results are presented by the symbols in Fig. 17.2b. The value of p ranges from 0 to 1, for it describes the relative area fraction of particles A with particles B. The maximum value of p_{a+b} is $\pi/4$ (≈ 0.79), for a circle cannot fill the whole square. Again, the infinite-matrix approximation is well behaved with both p ranging from 0 to 1 and p_{a+b} ranging from 0 to 0.79. An intuitive reason is that the underestimation of high thermal conductivity and the overestimation of low thermal conductivity, as discussed in Fig. 17.2a, cancel each other out when particles are big, e.g., $p_{a+b} = 0.79$.

Based on the results in Fig. 17.2, we feel it necessary to reunderstand the two famous effective medium theories, i.e., the Maxwell–Garnett formula and the Bruggeman formula. In terms of mathematical form, Eq. (17.11) is the generalized Maxwell–Garnett formula [26], and Eq. (17.12) is the generalized Bruggeman formula [27], by considering particle anisotropy. However, in this chapter, the understanding of Eqs. (17.11), (17.12) is totally different from the previous opinions [26–28] on three perspectives. (I) Random versus periodic: It is accepted that the two formulas can only explain completely random systems. However, the results in Fig. 17.2 illustrate that they are also well behaved in explaining periodic systems. (II) Small versus big: The Maxwell–Garnett formula is thought to be applicative only when particles are small enough. However, the results in Fig. 17.2a demonstrate that the Maxwell–Garnett formula is still contributing when particles are big. (III) Independent versus dependent: The two formulas are believed to be independent to describe different systems. However, we unify the two previously-thought independent formulas together, and clarify that the Bruggeman formula may be a direct result of the Maxwell–Garnett formula, i.e., from Eqs. (17.11) to (17.12). Anyway, periodicity is the key to understand these results. On one hand, it helps to simplify the calculations to the two basic structures presented in Fig. 17.1c, d. On the other hand, it avoids the overlap between particles, especially when area fractions are big.

17.4   Finite-Element Simulation and Laboratory Experiment for Thermal Transparency

Now we are in the position to present the finite-element simulation and laboratory experiment of thermal transparency by tailing the periodic interparticle interaction between particles A and particles B with Eq. (17.12).

Figure 17.3 shows the finite-element simulation of thermal transparency. The thermal conductivity of the region excluding particles A and particles B are set as required

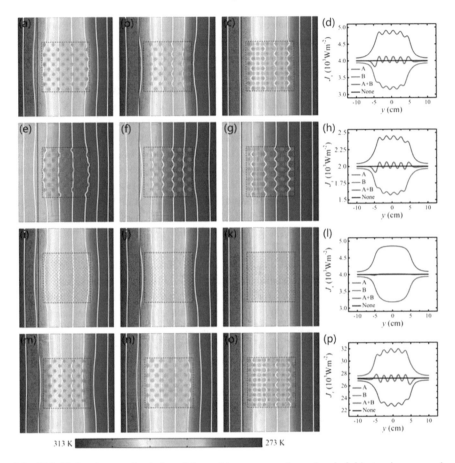

Fig. 17.3 Finite-element simulation of thermal transparency for **a–d**, **e–h**, **i–l** isotropic cases and **m–p** anisotropic case. κ_m is set as required by Eq. (17.12). The size of the simulation box is 20×20 cm, and that of the periodic composite material is 10×10 cm. The solid line is located at $x = -6$ cm, and the origin is located in the center of the simulation box. White lines represent isotherms. All particles are presented with 50% opacity in simulation boxes (particle color does not correspond to temperature). Blue, green, red, and black lines in **d**, **h**, **l**, **p** are the results with only particles A, only particles B, particles A plus particles B, and no particle, respectively. The thermal gradients in **a–d**, **i–l**, **m–p** and **e–h** are 2 K/cm and 1 K/cm, respectively. Other parameters: **a–d** $d_0 = 1$ cm, $\kappa_a = 400\,\text{Wm}^{-1}\text{K}^{-1}$, $\kappa_b = 1\,\text{Wm}^{-1}\text{K}^{-1}$, $p_a = 0.25$, $p_b = 0.25$, and $\kappa_m = 20\,\text{Wm}^{-1}\text{K}^{-1}$; **e–h** are the same as **a–d** except for the thermal gradient. **i–l** are the same as **a–d** except for $d_0 = 1/3$ cm; **m–p** $d_0 = 1$ cm, $\bar{\kappa}_a = \text{diag}(2, 0.5)\,\text{Wm}^{-1}\text{K}^{-1}$, $\kappa_b = 400\,\text{Wm}^{-1}\text{K}^{-1}$, $p_a = 0.18$, $p_b = 0.34$, $t/s = 1.2$, and $\kappa_m = 136\,\text{Wm}^{-1}\text{K}^{-1}$. Adapted from Ref. [29]

by Eq. (17.12). We only put particles A into the background, and high thermal conductivity repels isotherms; see Fig. 17.3a. We only put particles B into the background, and low thermal conductivity attracts isotherms; see Fig. 17.3b. We put both particles A and particles B into the background, and thermal transparency is achieved with the periodic interparticle interaction between particles A and particles B; see Fig. 17.3c.

The heat fluxes on the solid black lines, J_s, in Fig. 17.3a–c are presented in Fig. 17.3d. The red line (with some fluctuation) echoes with the black line in Fig. 17.3d, which validates the existence of thermal transparency. Then we only change the thermal gradient; see Fig. 17.3e–g. The shapes of isotherms are exactly the same as those in Fig. 17.3a–c, although the concrete temperature and heat flux (Fig. 17.3h) are different. In fact, the thermal gradient can be any value because the thermal property determined by Eq. (17.12) is independent of the thermal gradient. We further change the parameter d_0 and keep other parameters unchanged; see Fig. 17.3i–l. The results are basically the same as those in Fig. 17.3a–d, but with smaller fluctuation. Therefore, d_0 dose not affect the κ_m predicted by Eq. (17.12), but affects fluctuation. Finally we discuss an anisotropic case with material anisotropy of particle A and geometry anisotropy of particle B; see Fig. 17.3m–p. Anisotropic particle A with low thermal conductivity attracts isotherms, whereas anisotropic particle B with high thermal conductivity repels isotherms. Thermal transparency is still obtained when particles A and particles B are put together; see Fig. 17.3o. Hence the red line (with some fluctuation) agrees with the black line as shown in Fig. 17.3p.

Figure 17.4 shows the laboratory experiment of thermal transparency. For experimental realization, κ_a, κ_b, and κ_m are three related thermal conductivities, which are all set to be isotropic for feasibility. We use copper and air as the material of particle A and particle B, respectively. κ_m is determined by Eq. (17.12), and realized by carving air holes on a copper plate with effective medium theory [8]. Therefore, we only use two materials (copper and air) to realize three thermal conductivities (κ_a, κ_b, and κ_m). Note that the background possesses microstructures which is composed of copper (matrix) and air holes (microstructures) with area fraction 0.33. The three samples are fabricated by laser cutting, and measured between hot and cold baths by using the infrared camera (FLIR E60); see Fig. 17.4a–c. The upper and lower surfaces are respectively covered with transparent plastic and foamed plastic (both insulated) to reduce the influence of infrared reflection and thermal convection. A real product of Fig. 17.4c is presented in the Supplementary Proof; see Fig. 17.5. The measured results are presented in Fig. 17.4d–f. We also perform finite-element simulations based on the three samples, and the results are presented in Fig. 17.4g–i.

Moreover, we directly set $\kappa_m = 200\,\mathrm{Wm^{-1}K^{-1}}$ to perform simulation, rather than the structure of a copper plate carved with air holes, and the corresponding results are presented in Fig. 17.4j–l. The accordance between the laboratory experiment (Fig. 17.4d–f) and finite-element simulation (Fig. 17.4g–i or Fig. 17.4j–l) validates the scheme of tailoring periodic interparticle interaction for achieving thermal transparency.

According to results presented Fig. 17.4i, we also change the thermal gradient, matrix thermal conductivity, and matrix area fraction to study the changes of thermal transparency (particles A and particles B are kept unchanged). The results are presented in the Supplementary Proof; see Fig. 17.6, which conclude that the thermal gradient will not affect thermal transparency, but matrix properties will affect. The physical understanding is that the dominant Eq. (17.12) is independent of the thermal gradient, but dependent on the matrix properties.

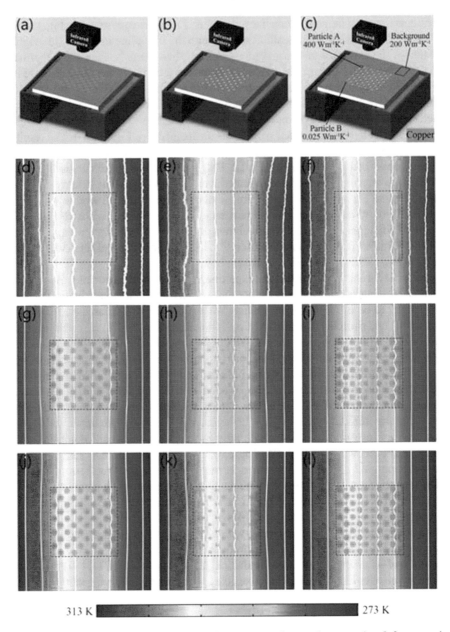

Fig. 17.4 Laboratory experiments for thermal transparency for **a–c** three samples, **d–f** measured results, **g–i** corresponding finite-element simulation results, and **j–l** finite-element simulations by directly setting $\kappa_m = 200\,\mathrm{Wm^{-1}K^{-1}}$. The sample size is $20 \times 20\,\mathrm{cm}$, and the size of the periodic composite material is $10 \times 10\,\mathrm{cm}$. Hot and cold baths are set at $313\,\mathrm{K}$ and $273\,\mathrm{K}$, respectively. The thermal conductivities of copper and air are 400 and $0.025\,\mathrm{Wm^{-1}K^{-1}}$, respectively. $\kappa_a = 400\,\mathrm{Wm^{-1}K^{-1}}$, $\kappa_b = 0.025\,\mathrm{Wm^{-1}K^{-1}}$, $d_0 = 1\,\mathrm{cm}$, $p_a = 0.33$, and $p_b = 0.11$. The area fraction of air holes (excluding particles A and particles B) in **a–c** is 0.33, and the distance between these air holes is $0.25\,\mathrm{cm}$. All particles are presented with 50% opacity in simulation boxes except for experimental results (particle color does not correspond to temperature). Adapted from Ref. [29]

17.5 Discussion and Conclusion

In this chapter, we have investigated the periodic interparticle interaction between particles A and particles B, which are commutable. However, this is just a beginning for exploring periodic interparticle interaction, and deeper mechanisms remain to be studied, at least from two aspects. (I) Lattice type: Here we arrange the particles with the simplest square lattice; see Fig. 17.1a. A question is how to calculate other lattice types. For example, a rectangle lattice is sure to bring anisotropy compared with the square lattice, let alone the more complicated orthorhombic lattice. (II) Relative position: Here the particles are alternately arranged along both horizontal and vertical directions; see Fig. 17.1a. A question is how to calculate other relative positions. For example, if the particles are alternately arranged only along horizontal direction, the new relative position will also bring anisotropy.

In summary, we have proposed a distinct mechanism for achieving thermal transparency which is induced by the symmetric interaction between periodic particles, namely periodic interparticle interaction, rather than neutral inclusions [4]. Theoretical analysis, finite-element simulations, and laboratory experiments all validate the proposed mechanism. Moreover, we reunderstand the Maxwell–Garnett formula and the Bruggeman formula from three aspects. That is, the two formulas can explain (I) periodic systems (II) with large area fractions of particles, and (III) they depend on each other. Our mechanism is feasible to handle many-particle systems for removing thermal stress concentration, preventing infrared detection, etc. On the same footing, periodic interparticle interaction proposed in this chapter can also be extended to realize other functions, such as thermal camouflaging.

17.6 Supplementary Proof

To realize the three-dimensional thermal transparency, Eqs. (17.5), (17.8), (17.10) correspondingly become

$$\langle J_a' \rangle = \kappa_a' \langle E_a' \rangle = v'\kappa_{\rho\rho}' \frac{3E_0'\kappa_m'}{v'\kappa_{\rho\rho}' + 2\kappa_m'}, \tag{17.13}$$

$$\langle J_b' \rangle = \kappa_b' \langle E_b' \rangle = \kappa_b' \frac{E_0'\kappa_m'}{\kappa_b'L' + \kappa_m'(1 - L')}, \tag{17.14}$$

$$\langle J_m' \rangle = \kappa_m' \langle E_m' \rangle = \kappa_m' E_0', \tag{17.15}$$

where the symbols ′ represent the case of three dimensions. $\kappa_a'\ (= v'\kappa_{\rho\rho}')$ is the effective scalar thermal conductivity of the particle with tensorial $\bar{\kappa}_a'\ [= \mathrm{diag}\,(\kappa_{\rho\rho}',\ \kappa_{\theta\theta}',\ \kappa_{\varphi\varphi}')$ with $\kappa_{\theta\theta}' = \kappa_{\varphi\varphi}']$, and $v' = -1/2 + \sqrt{1/4 + 2\kappa_{\theta\theta}'/\kappa_{\rho\rho}'}$. The horizontal shape factor L' can be calculated by

$$L' = \frac{s't'w'}{2} \int_0^\infty \frac{du}{\left(t'^2 + u\right)\sqrt{\left(s'^2 + u\right)\left(t'^2 + u\right)\left(w'^2 + u\right)}}, \qquad (17.16)$$

where w' is the third semi axis of the ellipsoid.

Owing to Eqs. (17.13), (17.14), (17.15), (17.11), (17.12) should respectively become

$$\kappa'_e = \frac{p'_a\kappa'_a\langle E'_a\rangle + p'_b\kappa'_b\langle E'_b\rangle + p'_m\kappa'_m\langle E'_m\rangle}{p'_a\langle E'_a\rangle + p'_b\langle E'_b\rangle + p'_m\langle E'_m\rangle} = \frac{p'_a\eta'_a\kappa'_a + p'_b\eta'_b\kappa'_b + p'_m\kappa'_m}{p'_a\eta'_a + p'_b\eta'_b + p'_m}, \quad (17.17)$$

$$\kappa'_m = \frac{-B' + \sqrt{B'^2 - 4A'C'}}{2A'}, \qquad (17.18)$$

where $p'_a = 2\pi r'^3/\left(3d_0'^3\right)$, $p'_b = 2\pi s't'w'/\left(3d_0'^3\right)$, $p'_m = 1 - p'_a - p'_b$, $\eta'_a = \langle E'_a\rangle/\langle E'_m\rangle = 3\kappa'_m/\left(\kappa'_a + 2\kappa'_m\right)$, $\eta'_b = \langle E'_b\rangle/\langle E'_m\rangle = \kappa'_m/\left[\kappa'_bL' + \kappa'_m\left(1 - L'\right)\right]$, $A' = p'\left(1 - 3L'\right) + 2$, $B' = -\kappa'_a\left[p'\left(4 - 3L'\right) - 1\right] + \kappa'_b\left[p'\left(2 + 3L'\right) - 2\right]$, $C' = \kappa'_a\kappa'_b\left[p'\left(1 - 3L'\right) - 1\right]$, and $p' = p'_a/\left(p'_a + p'_b\right)$. The κ'_m determined by Eq. (17.18) can describe the three-dimensional periodic interparticle interaction, because the right-hand side of Eq. (17.18) is independent of the matrix property.

The real product of Fig. 17.4c is presented in Fig. 17.5.

According to result in Fig. 17.4i, we keep particles A and particles B unchanged, and change the thermal gradient and matrix properties to study the changes of thermal transparency. We change the thermal gradient to be 1 K/cm (Fig. 17.6a) and 3 K/cm (Fig. 17.6b), and thermal transparency keeps unchanged. We change the matrix thermal conductivity from 400 to 1000 Wm^{-1}K^{-1} (Fig. 17.6c) and 100 Wm^{-1}K^{-1} (Fig. 17.6d). The isotherms are attracted and repelled, respectively. Finally, we change the area fraction of the voids from 0.33 to 0.05 (Fig. 17.6e) and 0.66 (Fig. 17.6f). The isotherms are also attracted and repelled, respectively.

17.7   Exercises and Solutions

Exercises

1. Consider a two-dimensional ellipse with major semi-axis $s = 1$ and minor semi-axis $t = 2$. Please calculate the shape factors along the major and minor axes.

Solutions

1. **Solution:** For major axis, $L = \dfrac{st}{2}\displaystyle\int_0^\infty \frac{du}{\left(t^2 + u\right)\sqrt{\left(s^2 + u\right)\left(t^2 + u\right)}} = t/\left(s + t\right)$

$= 2/3$. For minor axis, $L = \dfrac{st}{2}\displaystyle\int_0^\infty \frac{du}{\left(s^2 + u\right)\sqrt{\left(s^2 + u\right)\left(t^2 + u\right)}} = s/\left(s + t\right) =$

$1/3$.

Fig. 17.5 Real product of Fig. 17.4c with **a** side view, **b** top view, and **c** bottom view. Adapted from Ref. [29]

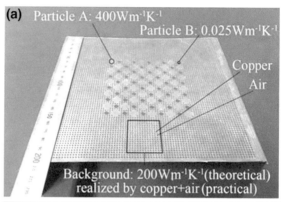

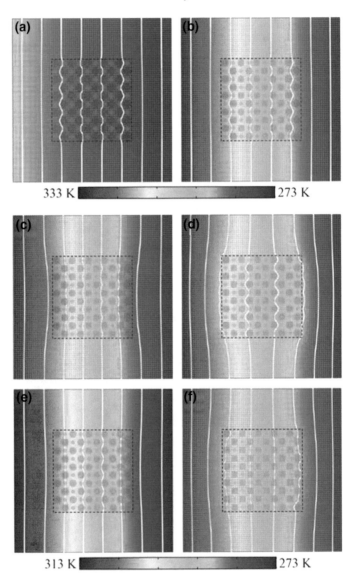

Fig. 17.6 Simulations results of other conditions. The parameters are the same as those in Fig. 17.4i except for **a** thermal gradient is 1 K/cm; **b** thermal gradient is 3 K/cm; **c** matrix thermal conductivity is $1000\,\mathrm{Wm^{-1}K^{-1}}$; **d** matrix thermal conductivity is $100\,\mathrm{Wm^{-1}K^{-1}}$; **e** area fraction of the voids is 0.05; and **f** area fraction of the voids is 0.66. Adapted from Ref. [29]

References

1. Fan, C.Z., Gao, Y., Huang, J.P.: Shaped graded materials with an apparent negative thermal conductivity. Appl. Phys. Lett. **92**, 251907 (2008)
2. Chen, T.Y., Weng, C.N., Chen, J.S.: Cloak for curvilinearly anisotropic media in conduction. Appl. Phys. Lett. **93**, 114103 (2008)
3. Maldovan, M.: Sound and heat revolutions in phononics. Nature **503**, 209–217 (2013)
4. He, X., Wu, L.Z.: Thermal transparency with the concept of neutral inclusion. Phys. Rev. E **88**, 033201 (2013)
5. Zeng, L.W., Song, R.X.: Experimental observation of heat transparency. Appl. Phys. Lett. **104**, 201905 (2014)
6. Yang, T.Z., Bai, X., Gao, D.L., Wu, L.Z., Li, B.W., Thong, J.T.L., Qiu, C.W.: Invisible sensors: simultaneous sensing and camouflaging in multiphysical fields. Adv. Mater. **27**, 7752–7758 (2015)
7. Wang, R.Z., Xu, L.J., Ji, Q., Huang, J.P.: A thermal theory for unifying and designing transparency, concentrating and cloaking. J. Appl. Phys. **123**, 115117 (2018)
8. Xu, L.J., Jiang, C.R., Shang, J., Wang, R.Z., Huang, J.P.: Periodic composites: quasi-uniform heat conduction, Janus thermal illusion, and illusion thermal diodes. Eur. Phys. J. B **90**, 221 (2017)
9. Shang, J., Jiang, C.R., Xu, L.J., Huang, J.P.: Many-particle thermal invisibility and diode from effective media. J. Heat Transfer **140**, 092004 (2018)
10. Narayana, S., Sato, Y.: Heat flux manipulation with engineered thermal materials. Phys. Rev. Lett. **108**, 214303 (2012)
11. Schittny, R., Kadic, M., Guenneau, S., Wegener, M.: Experiments on transformation thermodynamics: molding the flow of heat. Phys. Rev. Lett. **110**, 195901 (2013)
12. Xu, H.Y., Shi, X.H., Gao, F., Sun, H.D., Zhang, B.L.: Ultrathin three-dimensional thermal cloak. Phys. Rev. Lett. **112**, 054301 (2014)
13. Han, T.C., Bai, X., Gao, D.L., Thong, J.T.L., Li, B.W., Qiu, C.-W.: Experimental demonstration of a bilayer thermal cloak. Phys. Rev. Lett. **112**, 054302 (2014)
14. Ma, Y.G., Liu, Y.C., Raza, M., Wang, Y.D., He, S.L.: Experimental demonstration of a multiphysics cloak: manipulating heat flux and electric current simultaneously. Phys. Rev. Lett. **113**, 205501 (2014)
15. Kapadia, R.S., Bandaru, P.R.: Heat flux concentration through polymeric thermal lenses. Appl. Phys. Lett. **105**, 233903 (2014)
16. Xu, L.J., Yang, S., Huang, J.P.: Thermal theory for heterogeneously architected structure: fundamentals and application. Phys. Rev. E **98**, 052128 (2018)
17. Han, T.C., Bai, X., Thong, J.T.L., Li, B.W., Qiu, C.-W.: Full control and manipulation of heat signatures: cloaking, camouflage and thermal metamaterials. Adv. Mat. **26**, 1731–1734 (2014)
18. He, X., Wu, L.Z.: Illusion thermodynamics: a camouflage technique changing an object into another one with arbitrary cross section. Appl. Phys. Lett. **105**, 221904 (2014)
19. Yang, T.Z., Su, Y.S., Xu, W.K., Yang, X.D.: Transient thermal camouflage and heat signature control. Appl. Phys. Lett. **109**, 121905 (2016)
20. Hu, R., Zhou, S.L., Li, Y., Lei, D.Y., Luo, X.B., Qiu, C.W.: Illusion thermotics. Adv. Mater. **30**, 1707237 (2018)
21. Zhou, S.L., Hu, R., Luo, X.B.: Thermal illusion with twinborn-like heat signatures. Int. J. Heat Mass Transfer **127**, 607 (2018)
22. Xu, L.J., Wang, R.Z., Huang, J.P.: Camouflage thermotics: a cavity without disturbing heat signatures outside. J. Appl. Phys. **123**, 245111 (2018)
23. Xu, L.J., Huang, J.P.: A transformation theory for camouflaging arbitrary heat sources. Phys. Lett. A **382**, 3313 (2018)
24. Xu, L.J., Yang, S., Huang, J.P.: Designing the effective thermal conductivity of materials of core-shell structure: theory and simulation. Phys. Rev. E **99**, 022107 (2019)
25. Han, T.C., Yang, P., Li, Y., Lei, D.Y., Li, B.W., Hippalgaonkar, K., Qiu, W.: Full-parameter omnidirectional thermal metadevices of anisotropic geometry. Adv. Mater. **30**, 1804019 (2018)

26. Garnett, J.C.M.: Colours in metal glasses and in metallic films. Philos. Trans. R. Soc. Lond. Ser. A **203**, 385 (1904)
27. Bruggeman, D.A.G.: Berechnung verschiedener physikalischer Konstanten von heterogenen substanzen. I. Dielektrizitätskonstanten und Leitfähigkeiten der Mischkörper aus isotropen Substanzen (Calculation of different physical constants of heterogeneous substances. I. Dielectricity and conductivity of mixtures of isotropic substances). Annalen der Physik **24**, 636–664 (1935)
28. Wang, M., Pan, N.: Predictions of effective physical properties of complex multiphase materials. Mat. Sci. Eng. R **63**, 1 (2008)
29. Xu, L.J., Yang, S., Huang, J.P.: Thermal transparency induced by periodic interparticle interaction. Phys. Rev. Appl. **11**, 034056 (2019)

Chapter 18
Theory with Uniqueness Theorem: Thermal Camouflage

Abstract Cloaks can protect objects without disturbing heat signatures outside, and hence objects are invisible to the external detection. However, cloaks themselves are visible to inside detection because they possess different heat signatures from the outside. This fact limits applications. This problem is solved by developing a different theory in thermotics, and then a scheme of thermal supercavity is proposed, which is a cavity without disturbing heat signatures outside. Then different supercavities are introduced with various shapes in two or three dimensions, and they are also validated by using simulations and experiments. A scheme of super-invisibility is further designed, which makes the cavity itself also invisible to inside detection. Moreover, the scheme simplifies the complicated parameters of non-circle shaped cloaks, which requires only two natural materials with simple layer structure. This chapter is useful for achieving new kinds of thermal devices including thermal camouflage and designing similar supercavities in magnetostatics, electrostatics, particle diffusion, etc.

Keywords Uniqueness theorem · Thermal camouflage · Thermal supercavity · Super-invisibility · Cloak

18.1 Opening Remarks

Invisibility is a long-standing dream of human beings. Since the theory of transformation optics was put forward [1, 2], electromagnetic invisibility has attracted much attention (e.g., see Refs. [3–5]). In the duration, the physical fields have also been extended from those described by wave equations (say, in electromagnetics/optics [1–5] and acoustics [6, 7]) to those determined by diffusion equations (e.g., in thermotics [8–10]). Furthermore, the theories have been developed from the original transformation optics to others (e.g., directly solving the Laplace equation [11–20]), in order to design practical structures.

However, the existing thermal cloaks [12–14, 21, 22] are faced with a common problem: cloaks themselves are visible to inside detection. Let us take a thermal cloak (designed by transformation thermotics) as an example; see Fig. 18.1a. Thermal

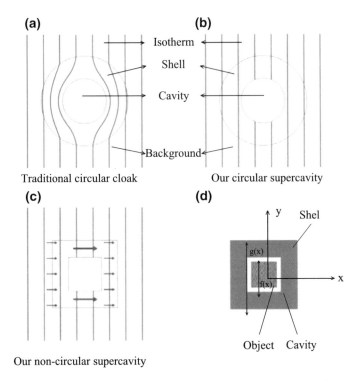

Fig. 18.1 Schematic diagram showing our concept: **a** the traditional circular cloak contains a distorted temperature field in the shell, thus causing the shell to be "visible" to inside thermal detection; **b** our circular supercavity proposed in this chapter has an undistorted temperature field in the shell, thus yielding the shell to be "invisible". Besides the circular case depicted in (**b**), our supercavity can be non-circular as well: **c** shows an example of square supercavity, whose details are indicated in (**d**). Brown lines and red arrows represent isotherms and heat flux, respectively. Adapted from Ref. [23]

cloak (a shell surrounding the cavity) can guide heat to flow around the cavity (in which arbitrary objects can be placed) without disturbing heat signatures outside the cloak (or in the background). Thus, objects located in the cavity are "invisible" to outside detection (Fig. 18.1a). However, thermal cloaks themselves are "visible" to inside detection (Fig. 18.1a) since they have different heat signatures from the outside as determined by the associated theories (say, the theory of transformation thermotics) [21, 22]. This fact limits applications, e.g., in the cases of misleading infrared detection.

To solve this problem, here we establish a different theory, which allows us to propose the scheme of supercavity as schematically shown in Fig. 18.1b, c. Clearly, the cavity does not disturb external heat signatures including those in the region of shell.

We further design the super-invisibility which makes both shell and cavity invisible to inside detections. Therefore, as pioneered by the work in Ref. [16], the whole system is invisible to whatever detections, which will contribute to thermal camouflage or illusion [15, 16, 24–26].

The aforementioned behaviors will be confirmed by simulations and experiments in this chapter. To proceed, let us first present the theory.

18.2 Theory Based on Uniqueness Theorem

We start by presenting the Fourier law that governs the process of heat conduction,

$$J = -\kappa \nabla T, \tag{18.1}$$

where J, κ, and T are respectively heat flux, thermal conductivity, and temperature.

As shown in Fig. 18.1b–d, our system contains three parts: shell (constructed by anisotropic materials in this chapter), cavity (in which arbitrary objects can be placed), and background. Comparing with traditional cloaks where temperature fields are distorted in the shell (Fig. 18.1a), we expect to keep the temperature fields of both shell and background the same (Fig. 18.1b, c), no matter what kinds of objects are placed in the cavity. In fact, if we can artificially match the boundary conditions (temperature T and normal heat flux J) between the shell and background, our expectation could be achieved indeed. In what follows, we give a proof to verify our idea.

Let us consider the background with two solutions of temperatures (T' and T'') and heat fluxes (J' and J''), and introduce an auxiliary function $Z(r) = (T' - T'')(J' - J'')$. Supposing that we have artificially matched the boundary conditions between background and shell,

$$\begin{aligned} T'_{\Sigma} &= T''_{\Sigma} \\ (J' \cdot e_n)_{\Sigma} &= (J'' \cdot e_n)_{\Sigma}, \end{aligned} \tag{18.2}$$

where Σ represents the boundary, and e_n is the unit normal vector of the boundary. So the integral value of $Z(r)$ on the boundary Σ must be zero,

$$\int_{\Sigma} Z(r) \cdot ds = \int_{V} \nabla \cdot Z(r) \, d\tau = 0. \tag{18.3}$$

We then calculate the divergence of $Z(r)$,

$$\nabla \cdot Z(r) = (\nabla T' - \nabla T'') \cdot (J' - J'') + (T' - T'') \cdot (\nabla \cdot J' - \nabla \cdot J''). \tag{18.4}$$

For the same background, we obtain

$$\nabla \cdot J' = \nabla \cdot J'' = q, \tag{18.5}$$

where q is the energy generated per unit volume and per unit time. Then Eq. (18.4) becomes

$$\boldsymbol{\nabla} \cdot \boldsymbol{Z}(\boldsymbol{r}) = \left(\boldsymbol{\nabla} T' - \boldsymbol{\nabla} T''\right) \cdot \left(\boldsymbol{J}' - \boldsymbol{J}''\right). \tag{18.6}$$

And Eq. (18.3) reads

$$\int_V \left(\boldsymbol{\nabla} T' - \boldsymbol{\nabla} T''\right) \cdot \left(\boldsymbol{J}' - \boldsymbol{J}''\right) d\tau = 0. \tag{18.7}$$

We further reduce Eq. (18.7) according to Eq. (18.1),

$$-\int_V \kappa \left(\boldsymbol{\nabla} T' - \boldsymbol{\nabla} T''\right)^2 d\tau = 0. \tag{18.8}$$

Since κ is always positive, we achieve

$$\boldsymbol{\nabla} T' = \boldsymbol{\nabla} T''. \tag{18.9}$$

Then we can conclude that if one can match the boundary conditions between the background and shell, the temperature profile of background can remain unchanged. To match the boundary conditions between background and shell, we can design the temperature distribution as follows,

$$\begin{aligned} (\text{Gradient } T)_x &= \boldsymbol{\nabla} T_0, \\ (\text{Gradient } T)_y &= 0, \end{aligned} \tag{18.10}$$

where $(\text{Gradient } T)_x$ (or $(\text{Gradient } T)_y$) is the horizontal (or vertical) thermal gradient in the shell, and $\boldsymbol{\nabla} T_0$ is the thermal gradient in the background. We do not consider the boundary conditions of heat flux because the normal heat flux is zero; see red arrows in Fig. 18.1c.

Now, we need to design the parameters of the shell to satisfy the required temperature distribution [Eq. (18.10)]. According to the conservation of heat flux in the shell; see red arrows in Fig. 18.1c, we obtain

$$\begin{aligned} -\kappa_{xx} \left[g(x) - f(x)\right] | (\text{Gradient } T)_x | &= -\kappa_b g(x) |\boldsymbol{\nabla} T_0|, \\ -\kappa_{yy}| (\text{Gradient } T)_y | &= \delta, \end{aligned} \tag{18.11}$$

where δ should be a non-zero (finite) variable, and κ_{xx} (or κ_{yy}) is the horizontal (or vertical) thermal conductivity of the shell (or anisotropic material), and κ_b is the thermal conductivity of background. $f(x)$ and $g(x)$ describe the length of the cavity and shell at the position x respectively, which have been depicted in Fig. 18.1d. Then we derive the thermal conductivity κ_{ani1} of the shell (anisotropic material) as

$$\kappa_{ani1} = \begin{pmatrix} \frac{g(x)\kappa_b}{g(x)-f(x)} & 0 \\ 0 & \kappa_{yy} \end{pmatrix}, \tag{18.12}$$

where κ_{yy} should be ∞ to ensure a non-zero δ. Equation (18.12) just helps to design our desired shell which itself is also invisible; see Figs. 18.2, 18.3 and 18.4.

Moreover, our theory can even make the cavity invisible (say, super-invisibility) with some sacrifice; see Fig. 18.5. The cloak designed according to Eq. (18.12) works regardless of the change of objects in the cavity. However, if we expect to make the cavity also invisible, the shell should be designed according to the cavity (object). We can design as follows.

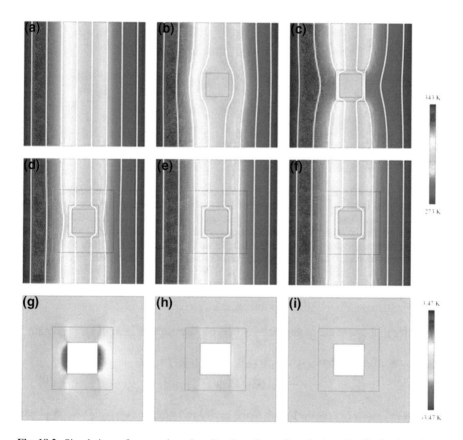

Fig. 18.2 Simulations of square shaped cavity: the color surface displays the distribution of temperature (**a–f**) or temperature difference (**g–i**), and the white lines in **a–f** represent the isotherms. **a** Background material (silica gel) with thermal conductivity 3.6 W/(m K) and size 8 × 8 cm; **b** The same background material whose central square area is occupied by an object (red copper) with 397 W/(m K) and 1.5 × 1.5 cm; **c** same as **b**, but the object is wrapped in air with 0.026 W/(m K) and 2 × 2 cm. **d–f** Are same as **c**, but the air is further wrapped in a square shell (that is constructed by an anisotropic material) with 4 × 4 cm. For the square shell (or the anisotropic material), the thermal conductivity is determined according to Eq. (18.12): $\kappa_{yy} = 20$ (**d**), 200 (**e**), and 2000 (**f**) W/(m K). The temperature difference between **d–f** and **a** is shown in **g–i**, respectively. In **g–i**, the central white square with 2 × 2 cm denotes the area involving both the object and air in **d–f**; this area just corresponds to the cavity as indicated in Fig. 18.1. Adapted from Ref. [23]

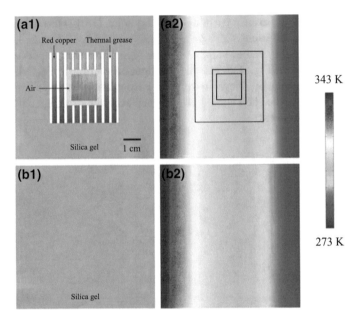

Fig. 18.3 Experiments of square shaped cavity: thermal images of experimental samples **a1** and **b1** are displayed in **a2** and **b2**, respectively. **a1** is the fabricated sample which is composed of a middle square (red copper) with size 1.5 × 1.5 cm; out of the middle square is air with 2 × 2 cm; out of the air is the shell composed of anisotropic material with 4 × 4 cm, which is made up of red copper and thermal grease; outermost is the background (silica gel) with 8 × 8 cm. **b1** Shows the sample of uniform background, namely, silica gel. Thermal conductivities of red copper, thermal grease, air, and silica gel are 397, 1.6, 0.026, and 3.6 W/(m K), respectively; the thickness of the two samples is about 0.3 mm. Adapted from Ref. [23]

Similar to the scheme of supercavity (the conservation of heat flux), we obtain

$$\{-\kappa_{xx}\left[n\left(x\right) - m\left(x\right)\right] - \kappa_c m\left(x\right)\} \mid (\text{Gradient } T)_x \mid = -\kappa_b n\left(x\right) \mid \nabla T_0 \mid,$$
$$-\kappa_{yy} \mid (\text{Gradient } T)_y \mid = \delta, \tag{18.13}$$

where $m\left(x\right)$ and $n\left(x\right)$ are the length of the cavity and shell, which have been depicted in Fig. 18.5a. Then we derive the thermal conductivity κ_{ani2} of the shell (or anisotropic material)

$$\kappa_{ani2} = \begin{pmatrix} \frac{n(x)\kappa_b - m(x)\kappa_c}{n(x) - m(x)} & 0 \\ 0 & \kappa_{yy} \end{pmatrix}, \tag{18.14}$$

where κ_{yy} should also be ∞, and κ_c is the thermal conductivity of cavity (object).

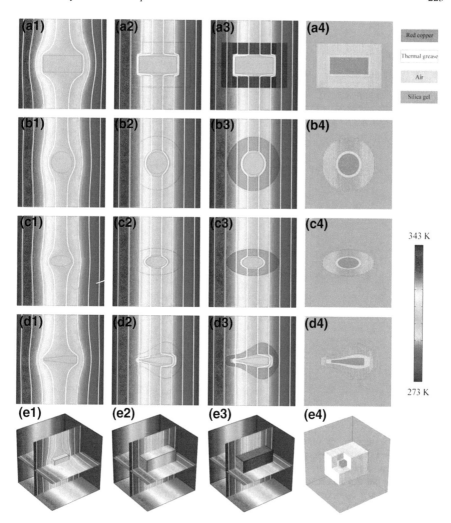

Red copper

Thermal grease

Air

Silica gel

343 K

273 K

Fig. 18.4 Simulations of various shaped cavities. **a1–a4** Show a rectangular case: **a1, a2** are respectively the same as Fig. 18.2b, f, but for the rectangular shape; for experimentally demonstrating **a2**, **a3** shows simulation results of the structure designed in (a4) according to Fig. 18.3 and Eq. (18.15). Similar to **a1–a4**, **b1–b4** show the circular shape, **c1–c4** the elliptic shape, **d1–d4** the irregular shape, and **e1–e4** the cuboid shape. For clarity, a small cuboid in the middle of **e4** is removed to display the inner structure. The results (namely, the uniform thermal gradients in shells) obtained from this figure are independent of sizes of the shells due to the generality of Eqs. (18.12) and (18.15). Thus here we omit such specific values of sizes adopted for simulation. Adapted from Ref. [23]

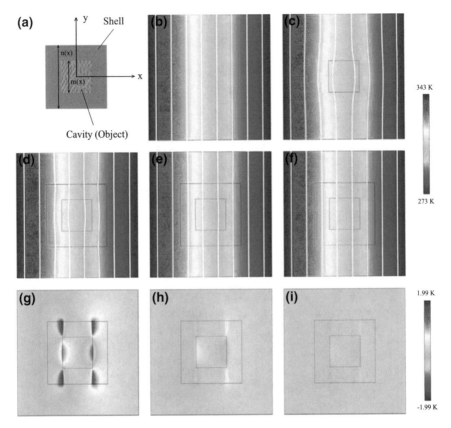

Fig. 18.5 Simulations of super-invisibility. **a** Schematic diagram of the design; **b** background material (silica gel) with thermal conductivity 3.6 W/(m K) and size 8 × 8 cm; **c** the same background material whose central square area is occupied by an object with 7 W/(m K) and 2 × 2 cm; **d**–**f** the same as (**b**), but the object is wrapped in a square shell (that is constructed by an anisotropic material) with 4 × 4 cm. For the square shell, the thermal conductivity is determined according to Eq. (18.14): $\kappa_{yy} = 20$ (**d**), 200 (**e**), and 2000 W/(m K) (**f**). The temperature difference between **d**–**f** and **b** is shown in **g**–**i**, respectively. Adapted from Ref. [23]

18.3 Simulations and Experiments of Square-Shaped Cavity

We perform finite-element simulations based on the commercial software COMSOL Multiphysics (http://www.comsol.com/) to show the validation of the aforementioned theory. Without loss of generality, we take square shape as an example, and display the simulations in Fig. 18.2a–f. Figure 18.2a shows the homogeneous background with uniform thermal field. Then we put an object (whose thermal conductivity is different from the background) into the background, and the temperature profile is distorted; see Fig. 18.2b. To remove the distortion, we add air outside the object to construct an insulation cavity (Fig. 18.2c). However, the thermal profile is still not recovered; see Fig. 18.2c. In Fig. 18.2d–f, three shells (anisotropic materials) with $\kappa_{yy} = 20$, 200 and 2000 W/(m K) according to Eq. (18.12) are applied to remove the temperature distortion. Figure 18.2d–f exhibits almost the same temperature profile as that in Fig. 18.2a; certainly larger κ_{yy} yields better comparison, which echoes with the prediction of Eq. (18.12). So the object is well hidden and the shell itself is also invisible. We further calculate the difference between Fig. 18.2a and d–f; see Fig. 18.2g–i. Clearly, the temperature distortion in the shell is well removed, especially for the case of larger κ_{yy}.

On the other hand, we also fabricated two samples (Fig. 18.3a1, b1) to verify the simulation results in Fig. 18.2. The anisotropic thermal conductivity of the shell, which is determined by Eq. (18.12), is designed with layer structures,

$$\kappa = \begin{pmatrix} \frac{(a/b+1)\kappa_a\kappa_b}{(a/b)\kappa_b+\kappa_a} & 0 \\ 0 & \frac{(a/b)\kappa_a+\kappa_b}{a/b+1} \end{pmatrix}, \tag{18.15}$$

where a (or b) is the length of uniform material with thermal conductivity κ_a (or κ_b).

Then we choose red copper and thermal grease with appropriate ratio according to Eq. (18.15) to fabricate the shell, and we utilize silica gel as the background material (Fig. 18.3a1). Figure 18.3a2 is the measured result of the sample shown in Fig. 18.3a1. Figure 18.3b1, b2 shows a reference group. Figure 18.3b2 is the experimental result of the homogeneous background shown in Fig. 18.3b1. Clearly the comparison between Fig. 18.3a2 and b2 is satisfactory. Small difference between experiments and simulations is caused by the thermal convection, and the experiments show the qualitative results.

18.4 Simulations of Various Shaped Cavities in Two or Three Dimensions

To show the robustness and generality of our theory, we further perform simulations for other shapes like rectangle, circle, ellipse, and irregular shape in two dimensions; we also investigate the shape of three-dimensional cuboid. See Fig. 18.4. In Fig. 18.4a1–e1, we place the objects with the shapes of rectangle, circle, ellipse, irregular shape and cuboid into the thermal field respectively. As a result, the temperature profile is undoubtedly distorted. Then we put these objects into the cavity associated with the shells [designed according to Eq. (18.12) where $\kappa_{yy} = 2000$ W/(m K)]; see Fig. 18.4a2–e2 that shows the desired effects (namely, the temperature gradient in the shell is the same as that in the background). Figure 18.4a3–e3 shows the simulation results of the practical structures displayed in Fig. 18.4a4–e4, respectively. The parameters of these structures are determined according to Eq. (18.15). The simulation results show that the shells themselves are invisible indeed while the objects are hidden in the central regions.

18.5   Simulations of Super-Invisibility

Similar simulations with Fig. 18.2 are also conducted to validate the super-invisibility; see Fig. 18.5. The cavity is fully filled with object, and there is no longer insulation material (air) between object and shell. Therefore, the shell is specific, which only works for certain object. Figure 18.5b shows a uniform thermal field in the homogeneous background. Then we put an object into the background; see Fig. 18.5c. In Fig. 18.5d–f, three shells with $\kappa_{yy} = 20, 200$ and 2000 W/(m K) according to Eq. (18.14) are applied to cancel out the temperature distortion. We also calculate the difference between Fig. 18.5b and d–f; see Fig. 18.5g–i. Clearly, the temperature distortion in the shell and cavity is both removed indeed, and hence super-invisibility is realized.

18.6   Discussion and Conclusion

Our supercavity is essentially a unidirectional passive cloak, for the boundary conditions are artificially matched. If the cavity is rotated or the uniform external field is changed, our scheme does not work again. Compared with the existing unidirectional active cloak [27–29], our scheme does not require extra sources, which is no doubt more applicable.

Our scheme also simplifies the extremely complicated parameters of non-circle shaped cloaks. As designed by transformation thermotics [10], non-circle shaped cloaks require extreme materials including inhomogeneity, anisotropy, and singularity, which are almost impossible to experimentally realize regardless of the development of thermal metamaterials. In contrast, our design requires only two natural materials with simple layer structure, which will bring great convenience and potential applications.

We have investigated the case of steady states only. Certainly, the unsteady states is subjected to further research because of the specific role of heat capacity [10, 30].

So far, we have established a theory and then proposed the scheme of thermal supercavity which makes the shell itself also invisible to inside detection. The effect has been confirmed in simulations and experiments. Only two natural materials (red copper and thermal grease, which are commercially available) were used to fabricate the shell in our experiment, which overcomes parameter complexity. Our theory is general for designing different shaped shells in both two and three dimensions. We also design the super-invisibility which makes the whole system invisible to whatever detections.

Our scheme enriches the research scale of thermal devices including thermal camouflage. It is useful for directly designing similar supercavity and super-invisibility in disciplines like magnetostatics [11], electrostatics [31] and particle diffusion [32], where electric conductivities, magnetic permeabilities and diffusion coefficients respectively play the same role as thermal conductivities in thermotics.

18.7   Exercises and Solutions

Exercises

1. Please prove the uniqueness theorem in theomotics. That is, if the sources, thermal conductivities, and boundary conditions are given, the thermal field is determined uniquely.

Solutions

1. **Solution**: The derivations are clearly shown in this chapter.

References

1. Pendry, J.B., Schurig, D., Smith, D.R.: Controlling electromagnetic fields. Science **312**, 1780–1782 (2006)
2. Leonhardt, U.: Optical conformal mapping. Science **312**, 1777–1780 (2006)
3. Schurig, D., Mock, J.J., Justice, B.J., Cummer, S.A., Pendry, J.B., Starr, A.F., Smith, D.R.: Metamaterial electromagnetic cloak at microwave frequencies. Science **314**, 977–980 (2006)
4. Liu, R., Ji, C., Mock, J., Chin, J.Y., Cui, T.J., Smith, D.R.: Broadband ground-plane cloak. Science **323**, 366–369 (2009)
5. Ergin, T., Stenger, N., Brenner, P., Pendry, J.B., Wegener, M.: Three-dimensional invisibility cloak at optical wavelengths. Science **328**, 337–339 (2010)
6. Chen, H.Y., Chan, C.T.: Acoustic cloaking in three dimensions using acoustic metamaterials. Appl. Phys. Lett. **91**, 183518 (2007)
7. Zhang, S., Xia, C., Fang, N.: Broadband acoustic cloak for ultrasound waves. Phys. Rev. Lett. **106**, 024301 (2011)
8. Fan, C.Z., Gao, Y., Huang, J.P.: Shaped graded materials with an apparent negative thermal conductivity. Appl. Phys. Lett. **92**, 251907 (2008)
9. Chen, T.Y., Weng, C.N., Chen, J.S.: Cloak for curvilinearly anisotropic media in conduction. Appl. Phys. Lett. **93**, 114103 (2008)
10. Guenneau, S., Amra, C., Veynante, D.: Transformation thermodynamics: cloaking and concentrating heat flux. Opt. Express **20**, 8207–8218 (2012)
11. Gomory, F., Solovyov, M., Souc, J., Navau, C., Camps, J.P., Sanchez, A.: Experimental realization of a magnetic cloak. Science **335**, 1466–1468 (2012)
12. Xu, H.Y., Shi, X.H., Gao, F., Sun, H.D., Zhang, B.L.: Ultrathin three-dimensional thermal cloak. Phys. Rev. Lett. **112**, 054301 (2014)
13. Han, T.C., Bai, X., Gao, D.L., Thong, J.T.L., Li, B.W., Qiu, C.-W.: Experimental demonstration of a bilayer thermal cloak. Phys. Rev. Lett. **112**, 054302 (2014)
14. Ma, Y.G., Liu, Y.C., Raza, M., Wang, Y.D., He, S.L.: Experimental demonstration of a multiphysics cloak: manipulating heat flux and electric current simultaneously. Phys. Rev. Lett. **113**, 205501 (2014)
15. Han, T.C., Bai, X., Thong, J.T.L., Li, B.W., Qiu, C.-W.: Full control and manipulation of heat signatures: cloaking, camouflage and thermal metamaterials. Adv. Mat. **26**, 1731–1734 (2014)
16. Yang, T.Z., Bai, X., Gao, D.L., Wu, L.Z., Li, B.W., Thong, J.T.L., Qiu, C.W.: Invisible sensors: simultaneous sensing and camouflaging in multiphysical fields. Adv. Mater. **27**, 7752–7758 (2015)
17. Yang, S., Xu, L.J., Wang, R.Z., Huang, J.P.: Full control of heat transfer in single-particle structural materials. Appl. Phys. Lett. **111**, 121908 (2017)

18. Wang, R.Z., Xu, L.J., Huang, J.P.: Thermal imitators with single directional invisibility. J. Appl. Phys. **122**, 215107 (2017)
19. Xu, L.J., Jiang, C.R., Shang, J., Wang, R.Z., Huang, J.P.: Periodic composites: quasi-uniform heat conduction, Janus thermal illusion, and illusion thermal diodes. Eur. Phys. J. B **90**, 221 (2017)
20. Wang, R.Z., Xu, L.J., Ji, Q., Huang, J.P.: A thermal theory for unifying and designing transparency, concentrating and cloaking. J. Appl. Phys. **123**, 115117 (2018)
21. Narayana, S., Sato, Y.: Heat flux manipulation with engineered thermal materials. Phys. Rev. Lett. **108**, 214303 (2012)
22. Schittny, R., Kadic, M., Guenneau, S., Wegener, M.: Experiments on transformation thermodynamics: molding the flow of heat. Phys. Rev. Lett. **110**, 195901 (2013)
23. Xu, L.J., Wang, R.Z., Huang, J.P.: Camouflage thermotics: a cavity without disturbing heat signatures outside. J. Appl. Phys. **123**, 245111 (2018)
24. Hu, R., Wei, X.L., Hu, J.Y., Luo, X.B.: Local heating realization by reverse thermal cloak. Sci. Rep. **4**, 3600 (2014)
25. Hu, R., Zhou, S.L., Li, Y., Lei, D.Y., Luo, X.B., Qiu, C.W.: Illusion thermotics. Adv. Mater. **30**, 1707237 (2018)
26. Li, Y., Bai, X., Yang, T.Z., Luo, H., Qiu, C.W.: Structured thermal surface for radiative camouflage. Nat. Commun. **9**, 273 (2018)
27. Ma, Q., Mei, Z.L., Zhu, S.K., Jin, T.Y., Cui, T.J.: Experiments on active cloaking and illusion for Laplace equation. Phys. Rev. Lett. **111**, 173901 (2013)
28. Nguyen, D.M., Xu, H.Y., Zhang, Y.M., Zhang, B.L.: Active thermal cloak. Appl. Phys. Lett. **107**, 121901 (2015)
29. Lan, C.W., Bi, K., Gao, Z.H., Li, B., Zhou, J.: Achieving bifunctional cloak via combination of passive and active schemes. Appl. Phys. Lett. **109**, 201903 (2016)
30. Yang, T.Z., Su, Y.S., Xu, W.K., Yang, X.D.: Transient thermal camouflage and heat signature control. Appl. Phys. Lett. **109**, 121905 (2016)
31. Yang, F., Mei, Z.L., Jin, T.Y., Cui, T.J.: DC electric invisibility cloak. Phys. Rev. Lett. **109**, 053902 (2012)
32. Guenneau, S., Puvirajesinghe, T.M.: Fick's second law transformed: one path to cloaking in mass diffusion. J. Roy. Soc. Interface **10**, 20130106 (2013)

Chapter 19
Theory for Thermal Radiation: Transparency, Cloak, and Expander

Abstract The existing thermal metamaterials are almost designed to work at room temperature where thermal conduction is the dominant way of heat transfer. Unfortunately, as the temperature increases, thermal radiation becomes more and more important, and hence these metamaterials no longer work. The inability to handle thermal radiation largely limits practical applications at high temperature, such as thermal protection. To solve this problem, here we describe an effective medium theory to manipulate thermal radiation with the Rosseland diffusion approximation. This theory helps to design three types of radiative metamaterials even with anisotropic geometries, including transparency, cloak, and expander. Theoretical analyses are further confirmed by finite-element simulations, which indicate that these radiative metamaterials perform well at both steady and transient states. This chapter not only introduces an effective medium theory to manipulate thermal radiation, but also designs three types of radiative metamaterials. These results may provide hints on novel thermal management and have potential applications in radiative illusion/camouflage, radiative diode, etc.

Keywords Radiative metamaterials · Rosseland diffusion approximation · Thermal transparency · Thermal cloak · Thermal expander

19.1 Opening Remarks

Since the proposal of transformation thermotics [1], thermal metamaterials have experienced prosperous developments and made abundant achievements. The capability of thermal management is largely improved with thermal metamaterials, such as thermal transparency [2–5], thermal cloak [6–11], thermal expander [12], etc. Thermal transparency aims to design a shell according to the inside core and ensure the temperature field outside the shell undistorted. Thermal cloak can protect any object inside it from being detected, which can be realized by setting the core in transparency scheme to be insulated. Thermal expander can efficiently enlarge a small source into a large one, which is based on two thermal cloaks. Moreover, the theory of transformation thermal convection [13–15] was also established, thus yielding

© Springer Nature Singapore Pte Ltd. 2020
J.-P. Huang, *Theoretical Thermotics*,
https://doi.org/10.1007/978-981-15-2301-4_19

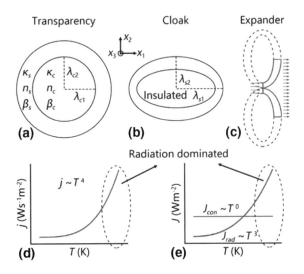

Fig. 19.1 Schematic diagrams of **a** thermal transparency, **b** thermal cloak, and **c** thermal expander. **d** and **e** Qualitatively show the radiative emittance j, conductive flux J_{con}, and radiative flux J_{rad} as a function of temperature T. Adapted from Ref. [39]

similar convective metamaterials. The revelation of the anti-parity-time symmetry in thermal convective systems [16] also largely promotes developments. These thermal metamaterials provide powerful methods to manipulate thermal conduction and thermal convection.

Unfortunately, none of these thermal metamaterials can work at high temperature where thermal radiation becomes the dominant way of heat transfer. In fact, thermal radiation is of great significance, which has attracted lots of research interest, such as radiative cooling [17–19], thermal memory [20, 21], etc. However, metamaterials designed for manipulating thermal radiation have been rarely touched due to the lack of appropriate theories. In consideration of practical applications like thermal protection at high temperature, it is urgent to establish a theory to manipulate thermal radiation efficiently.

Here, we propose an effective medium theory to design radiative metamaterials with the Rosseland diffusive approximation. By designing the parameters (mainly the thermal conductivity and Rosseland mean extinction coefficient) of a core-shell structure, we realize three types of radiative metamaterials even with anisotropic geometries, including transparency, cloak, and expander; see Fig. 19.1a–c. Finite-element simulations confirm our theoretical analyses, which indicate that these radiative metamaterials are well-behaved in both steady and transient states. In what follows, let us proceed with theory.

19.2 Theory

We consider a passive and steady process of heat transfer, where the total heat flux $\boldsymbol{J}_{total}$ (composed of the conductive flux $\boldsymbol{J}_{con}$ and the radiative flux $\boldsymbol{J}_{rad}$) is divergency-free,

$$\nabla \cdot \boldsymbol{J}_{total} = \nabla \cdot (\boldsymbol{J}_{con} + \boldsymbol{J}_{rad}) = 0. \tag{19.1}$$

The conductive flux $\boldsymbol{J}_{con}$ is dominated by the Fourier law,

$$\boldsymbol{J}_{con} = -\kappa \nabla T, \tag{19.2}$$

where κ is the thermal conductivity. The radiative flux $\boldsymbol{J}_{rad}$ is given by the Rosseland diffusion approximation,

$$\boldsymbol{J}_{rad} = -\frac{16 n^2 \sigma T^3}{3\beta} \nabla T, \tag{19.3}$$

where n is the relative refraction index, β is the Rosseland mean extinction coefficient, and σ is the Stefan-Boltzmann constant $(= 5.67 \times 10^{-8}\,\mathrm{Wm^{-2}K^{-4}})$.

We consider a three-dimensional core-shell structure (Fig. 19.1a) which has a core with thermal conductivity κ_c, relative refraction index n_c, and Rosseland mean extinction coefficient β_c coated by a shell with corresponding parameters κ_s, n_s, and β_s. The subscript c (or s) indicates the core (or shell) throughout this chapter. The semi-axis lengths of the core and shell are λ_{ci} and λ_{si} ($i = 1, 2, 3$), respectively. We introduce ellipsoidal coordinates (ρ, ξ, η) to proceed,

$$\begin{cases} \frac{x^2}{\rho+\lambda_1^2} + \frac{y^2}{\rho+\lambda_2^2} + \frac{z^2}{\rho+\lambda_3^2} = 1 \text{ (confocal ellipsoids)} \\ \frac{x^2}{\xi+\lambda_1^2} + \frac{y^2}{\xi+\lambda_2^2} + \frac{z^2}{\xi+\lambda_3^2} = 1 \text{ (hyperboloids of one sheet)} \\ \frac{x^2}{\eta+\lambda_1^2} + \frac{y^2}{\eta+\lambda_2^2} + \frac{z^2}{\eta+\lambda_3^2} = 1 \text{ (hyperboloids of two sheets)} \end{cases}, \tag{19.4}$$

where λ_1, λ_2, and λ_3 are three constants, satisfying $\rho > -\lambda_1^2 > \xi > -\lambda_2^2 > \eta > -\lambda_3^2$. Then, the Laplace equation $\nabla \cdot (-\kappa \nabla T) = 0$ can be expressed as

$$\frac{\partial}{\partial \rho} \left[g(\rho) \frac{\partial T}{\partial \rho} \right] + \frac{g(\rho)}{\rho + \lambda_i^2} \frac{\partial T}{\partial \rho} = 0, \tag{19.5}$$

where $g(\rho) = \sqrt{(\rho + \lambda_1^2)(\rho + \lambda_2^2)(\rho + \lambda_3^2)}$. Equation (19.5) has a solution as

$$T = \left[u + v \int_0^\rho (\rho + \lambda_i^2)^{-1} g(\rho)^{-1}\, d\rho \right] x_i, \tag{19.6}$$

where u and v are arbitrary constants, and (x_1, x_2, x_3) denotes Cartesian coordinates. We define the temperatures of the core, shell, and background as T_c, T_s, and T_b, respectively. They can be expressed as

$$\begin{cases} T_c = u_c x_i \\ T_s = \left[u_s + v_s \int_{\rho_c}^{\rho} \left(\rho + \lambda_i^2 \right)^{-1} g\left(\rho \right)^{-1} d\rho \right] x_i \,, \\ T_b = u_b x_i \end{cases} \qquad (19.7)$$

where u_c, u_s, and v_s can be determined by boundary conditions. The exterior surfaces of the core and shell are denoted by ρ_c and ρ_s, where boundary conditions are given by continuities of temperatures and normal heat fluxes, thus yielding

$$\begin{cases} u_c = u_s \\ u_b = u_s + v_s \int_{\rho_c}^{\rho_s} \left(\rho + \lambda_i^2 \right)^{-1} g\left(\rho \right)^{-1} d\rho \\ u_c = 2 v_s \kappa_s \left(\kappa_c - \kappa_s \right)^{-1} g\left(\rho_c \right)^{-1} \\ u_b = 2 v_s \kappa_s \left(\kappa_{ei} - \kappa_s \right)^{-1} g\left(\rho_s \right)^{-1} \end{cases} \qquad (19.8)$$

where κ_{ei} is the effective thermal conductivity of the core-shell structure along the direction of x_i. Solving Eq. (19.8) can directly derive the expression of κ_{ei}. Nevertheless, it is a complex formula which requires simplification. Thus, we define the semi-axis lengths of the core λ_{ci} and shell λ_{si} as

$$\begin{cases} \lambda_{ci} = \sqrt{\lambda_i^2 + \rho_c} \\ \lambda_{si} = \sqrt{\lambda_i^2 + \rho_s} \end{cases} \qquad (19.9)$$

where $i = 1,\ 2,\ 3$. Thus, the volume fraction f can be expressed as

$$f = \lambda_{c1} \lambda_{c2} \lambda_{c3} / \left(\lambda_{s1} \lambda_{s2} \lambda_{s3} \right) = g\left(\rho_c \right) / g\left(\rho_s \right). \qquad (19.10)$$

We also define the shape factor d_{wi} along the direction of x_i as

$$d_{wi} = \frac{\lambda_{w1} \lambda_{w2} \lambda_{w3}}{2} \int_0^\infty \left(\alpha + \lambda_{wi}^2 \right)^{-1} \left[\left(\alpha + \lambda_{w1}^2 \right) \left(\alpha + \lambda_{w2}^2 \right) \left(\alpha + \lambda_{w3}^2 \right) \right]^{-1/2} d\alpha, \qquad (19.11)$$

where w can take c or s, representing the shape factor of the core or shell. Thus, we have

$$\int_{\rho_c}^{\rho_s} \left(\rho + \lambda_i^2 \right)^{-1} g\left(\rho \right)^{-1} d\rho = \int_{\rho_c}^\infty \left(\rho + \lambda_i^2 \right)^{-1} g\left(\rho \right)^{-1} d\rho - \int_{\rho_s}^\infty \left(\rho + \lambda_i^2 \right)^{-1} g\left(\rho \right)^{-1} d\rho$$
$$= 2 d_{ci} g\left(\rho_c \right)^{-1} - 2 d_{si} g\left(\rho_s \right)^{-1}. \qquad (19.12)$$

Finally, we can derive the brief expression for κ_{ei} as

$$\kappa_{ei} = \kappa_s \left[\frac{f \, (\kappa_c - \kappa_s)}{\kappa_s + (d_{ci} - f d_{si}) \, (\kappa_c - \kappa_s)} + 1 \right]. \qquad (19.13)$$

This is a standard result for calculating the effective thermal conductivity [22]. The shape factors satisfy the sum rule $d_{w1} + d_{w2} + d_{w3} = 1$. Then, the effective thermal conductivity of any core-shell structure can, in principle, be derived with Eq. (19.13). However, only when the core-shell structure is confocal or concentric, can Eq. (19.13) predict the effective thermal conductivity strictly. Moreover, Eq. (19.13) can also be reduced to handle the cylindrical (two-dimensional) cases by taking $\lambda_{w3} = \infty$, thus yielding $d_{w1} = \lambda_{w2}/(\lambda_{w1} + \lambda_{w2})$, $d_{w2} = \lambda_{w1}/(\lambda_{w1} + \lambda_{w2})$, and $d_{w3} = 0$ ($d_{w1} + d_{w2} + d_{w3} = 1$ is still satisfied).

Since thermal radiation with the Rosseland diffusion approximation has the similar equation form as thermal conduction [Eqs. (19.2) and (19.3)], the radiative coefficient (denoted by $\gamma = n^2/\beta$) can be similarly calculated by

$$\gamma_{ei} = \gamma_s \left[\frac{f \, (\gamma_c - \gamma_s)}{\gamma_s + (d_{ci} - f d_{si}) \, (\gamma_c - \gamma_s)} + 1 \right], \qquad (19.14)$$

where γ_{ei} is the effective radiative coefficient of the core-shell structure along the direction of x_i. Equations (19.13) and (19.14) can predict the effective thermal conductivity and effective radiative coefficient. The only requirement is to keep κ/γ of different regions as a constant, for realizing the same effect of conduction and radiation.

19.3 Finite-Element Simulations

Further, we perform finite-element simulations with COMSOL MULTIPHYSICS (http://www.comsol.com/) to validate the theoretical analyses. Without loss of generality, we consider two-dimensional cases, set the simulation box to 10×10 cm^2, set the relative refraction index of all regions to 1, and set the thermal conductivity and Rosseland mean extinction coefficient of the background to 1 Wm^{-1}K^{-1} and 100 m^{-1}, respectively. These parameters meet the requirement of optically thick media where the Rosseland diffusion approximation is reasonable.

The Stefan-Boltzmann law suggests that the radiative emittance j is proportional to T^4, as qualitatively shown in Fig. 19.1d. In the presence of a same temperature gradient, the conductive flux J_{con} is independent of concrete temperatures, whereas the radiative flux J_{rad} is proportional to T^3, as qualitatively presented in Fig. 19.1e. These qualitative analyses illustrate that thermal radiation is of great significance at high temperature. Thus, three types of temperature settings are applied in our finite-element simulations. (I) 273–313 K, indicating a small upper temperature limit where conduction (Con.) is dominant. (II) 273–673 K, indicating a medium upper

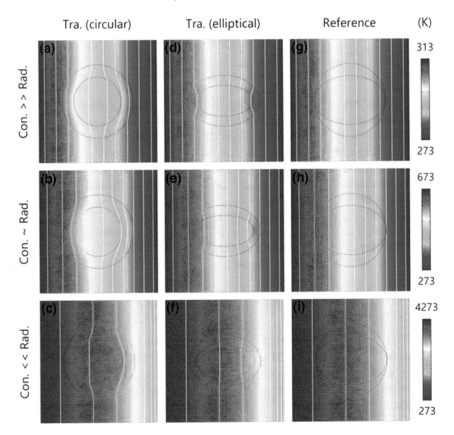

Fig. 19.2 Steady simulations of thermal transparency. **a–c**: $\lambda_{c1} = \lambda_{c2} = 2$ cm, $\lambda_{s1} = \lambda_{s2} = 3$ cm, $\kappa_c = 2$ Wm^{-1}K^{-1}, $\beta_c = 50$ m^{-1}, $\kappa_s = 0.62$ Wm^{-1}K^{-1}, and $\beta_s = 161.1$ m^{-1}. **d–f**: $\lambda_{c1} = 2.5$ cm, $\lambda_{c2} = 1.25$ cm, $\lambda_{s1} = 3$ cm, $\lambda_{s2} = 2.08$ cm, $\kappa_c = 0.5$ Wm^{-1}K^{-1}, $\beta_c = 200$ m^{-1}, $\kappa_s = 1.61$ Wm^{-1}K^{-1}, and $\beta_s = 62$ m^{-1}. **g–i** are references with pure background parameters. Circular (or elliptical) dashed lines are plotted for comparison with the circular (or elliptical) transparency. Adapted from Ref. [39]

temperature limit where conduction and radiation (Rad.) are roughly equal. (III) 273–4273 K, indicating a large upper temperature limit where radiation is dominant.

Thermal transparency (Tra.) is to design a shell according to the object, ensuring the temperature profile outside the shell undistorted; see Fig. 19.2. When parameters are delicately designed according to Eqs. (19.13) and (19.14), the temperature profile outside the shell is undistorted (Fig. 19.2a–c or 19.2d–f) as if there wasn't a core-shell structure in the center (Fig. 19.2g–i). With a small upper temperature limit where conduction is dominant (see the first row of Fig. 19.2), the temperature gradient outside the shell is uniform. With the increment of the upper temperature limit, radiation starts exerting an influence, thus yielding the nonuniform temperature gradients outside the shell (see the last two rows of Fig. 19.2).

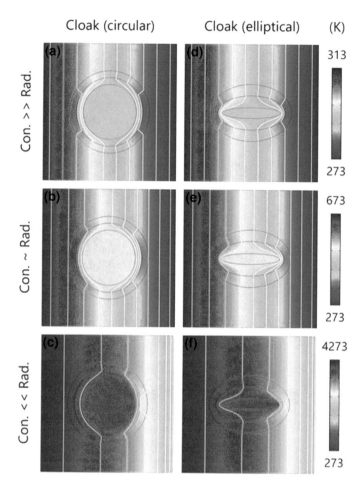

Fig. 19.3 Steady simulations of thermal cloaks. An inner object is coated by an insulated layer with $\kappa = 10^{-5}$ Wm^{-1}K^{-1} and $\beta = 10^5$ m^{-1}. Since the heat flux cannot enter into the insulated layer, the inner object plus the insulated layer can be equivalently regarded as an insulated core with $\kappa_c = 10^{-5}$ Wm^{-1}K^{-1} and $\beta_c = 10^5$ m^{-1}. Other parameters are as follows. **a–c**: $\lambda_{c1} = \lambda_{c2} = 2.5$ cm, $\lambda_{s1} = \lambda_{s2} = 3$ cm, $\kappa_s = 5.54$ Wm^{-1}K^{-1}, and $\beta_s = 18.1$ m^{-1}. **d–f**: $\lambda_{c1} = 2.5$ cm, $\lambda_{c2} = 1.25$ cm, $\lambda_{s1} = 3$ cm, $\lambda_{s2} = 2.08$ cm, $\kappa_s = 2.35$ Wm^{-1}K^{-1}, and $\beta_s = 42.5$ m^{-1}. Adapted from Ref. [39]

Thermal cloak can protect any object inside it from being detected, whose parameters are independent of the object. For this purpose, an insulated layer is required to keep the heat flux off the object. Then, the object plus the insulated layer can be equivalently regarded as an insulated core, say, $\kappa_c = \gamma = 0$. Further, we design a shell according to Eqs. (19.13) and (19.14) to remove the effect of the insulated core. The simulation results are presented in Fig. 19.3a–c and 19.3d–f. Clearly, the isotherms are kept off the object, indicating that the heat flux cannot enter into the object.

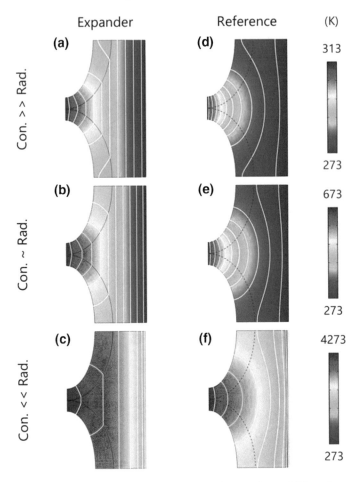

Fig. 19.4 Steady simulations of thermal expander. The sizes are $\lambda_{c1} = 2.08$ cm, $\lambda_{c2} = 4.17$ cm, $\lambda_{s1} = 3.46$ cm, $\lambda_{s2} = 5$ cm, and the width between hot and cold sources is 6 cm. Other parameters are as follows. **a–c**: $\kappa_s = 4.91$ Wm^{-1}K^{-1} and $\beta_s = 20.3$ m^{-1}. **d–f**: pure background parameters. Adapted from Ref. [39]

Thermal expander can efficiently enlarge a small source into a large one based on the design of two elliptical cloaks. Concretely speaking, we put two elliptical cloaks together, and take out a quarter of the whole structure as an expander; see Fig. 19.1c. As ensured by the uniqueness theorem in thermotics [23], the temperature distribution of the background isn't distorted, thus yielding the expander effect. Finite-element simulations are presented in Fig. 19.4a–c. Clearly, the isotherms of the background are straight, indicating a perfect performance. However, a pure background material strongly distorts the isotherms of the background; see Fig. 19.4d–f. Such device is flexible to adjust source sizes and has applications in uniform heating and effective dissipation.

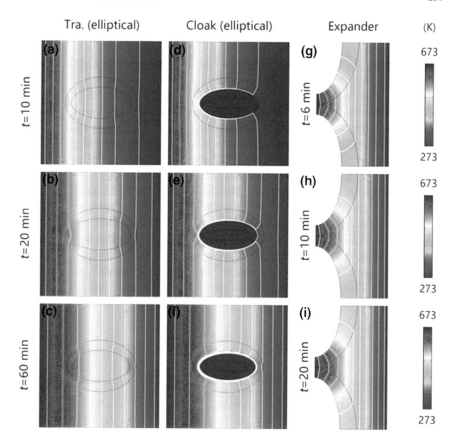

Fig. 19.5 Transient simulations of transparency, cloak, and expander. The sizes and material parameters of **a–c**, **d–f**, and **g–i** are the same as those of Figs. 19.2d–19.2f, 19.3d–19.3f, and 19.4a–19.4c, respectively. The density and heat capacity of the background are $\rho C = 10^6$ Jm^{-3}K^{-1}. Other parameters are as follows. **a–c**: $(\rho C)_c = 5 \times 10^5$ Jm^{-3}K^{-1} and $(\rho C)_s = 1.61 \times 10^6$ Jm^{-3}K^{-1}. **d–f**: $(\rho C)_s = 2.35 \times 10^6$ Jm^{-3}K^{-1}. **g–i**: $(\rho C)_s = 5 \times 10^5$ Jm^{-3}K^{-1}. Adapted from Ref. [39]

The above results only consider steady states. These radiative metamaterials can also be extended to transient states if we consider the density and heat capacity. For designing the transient transparency and cloak, we set the heat diffusivity $\kappa / (\rho C)$ to be a constant. Although it is an approximate method, its performance is still satisfying. The corresponding results at $t = 10$, 20, 60 mins are presented in Fig. 19.5a–19.5c and 19.5d–19.5f, respectively. For designing a transient expander, we use the optimization method and set the diffusivity of the shells larger than that of the background to achieve the best transient effect. The corresponding results at $t = 6$, 10, 20 mins are presented in Fig. 19.5g–i.

19.4   Discussion and Conclusion

In this chapter, thermal conduction is dealt with the Fourier law [Eq. (19.2)] which is an appropriate hypothesis at the macroscale. However, at the nanoscale, if the phonon effect is taken into consideration [24–28], the Fourier law may be invalid. Meanwhile, the thermal radiation under our consideration is handled with the Rosseland diffusion approximation [Eq. (19.3)] which requires the participating media to be optically thick. Namely, thermal radiation only propagates a short distance before being absorbed or scattered. Lots of other radiative models, such as those considering near-field radiation [29–32], remain to be further explored.

In summary, we have proposed an effective medium theory to manipulate thermal radiation and designed three types of metamaterials including transparency, cloak, and expander. All theoretical analyses are confirmed by finite-element simulations. These radiative metamaterials are well-behaved in both steady and transient states. Certainly, they have broad potential applications in designing thermal illusion/camouflage [33–35] and other thermal metamaterials [36–38] for the regime with high temperature where thermal radiation is the dominating effect.

References

1. Fan, C.Z., Gao, Y., Huang, J.P.: Shaped graded materials with an apparent negative thermal conductivity. Appl. Phys. Lett. **92**, 251907 (2008)
2. He, X., Wu, L.Z.: Thermal transparency with the concept of neutral inclusion. Phys. Rev. E **88**, 033201 (2013)
3. Zeng, L.W., Song, R.X.: Experimental observation of heat transparency. Appl. Phys. Lett. **104**, 201905 (2014)
4. Yang, T.Z., Bai, X., Gao, D.L., Wu, L.Z., Li, B.W., Thong, J.T.L., Qiu, C.W.: Invisible sensors: simultaneous sensing and camouflaging in multiphysical fields. Adv. Mater. **27**, 7752–7758 (2015)
5. Xu, L.J., Yang, S., Huang, J.P.: Thermal transparency induced by periodic interparticle interaction. Phys. Rev. Appl. **11**, 034056 (2019)
6. Narayana, S., Sato, Y.: Heat flux manipulation with engineered thermal materials. Phys. Rev. Lett. **108**, 214303 (2012)
7. Schittny, R., Kadic, M., Guenneau, S., Wegener, M.: Experiments on transformation thermodynamics: molding the flow of heat. Phys. Rev. Lett. **110**, 195901 (2013)
8. Xu, H.Y., Shi, X.H., Gao, F., Sun, H.D., Zhang, B.L.: Ultrathin three-dimensional thermal cloak. Phys. Rev. Lett. **112**, 054301 (2014)
9. Han, T.C., Bai, X., Gao, D.L., Thong, J.T.L., Li, B.W., Qiu, C.-W.: Experimental demonstration of a bilayer thermal cloak. Phys. Rev. Lett. **112**, 054302 (2014)
10. Ma, Y.G., Liu, Y.C., Raza, M., Wang, Y.D., He, S.L.: Experimental demonstration of a multiphysics cloak: manipulating heat flux and electric current simultaneously. Phys. Rev. Lett. **113**, 205501 (2014)
11. Li, Y., Zhu, K.J., Peng, Y.G., Li, W., Yang, T.Z., Xu, H.X., Chen, H., Zhu, X.F., Fan, S.H., Qiu, C.W.: Thermal meta-device in analogue of zero-index photonics. Nat. Mater. **18**, 48–54 (2019)
12. Han, T.C., Yang, P., Li, Y., Lei, D.Y., Li, B.W., Hippalgaonkar, K., Qiu, W.: Full-parameter omnidirectional thermal metadevices of anisotropic geometry. Adv. Mater. **30**, 1804019 (2018)
13. Guenneau, S., Petiteau, D., Zerrad, M., Amra, C., Puvirajesinghe, T.: Transformed Fourier and Fick equations for the control of heat and mass diffusion. AIP Adv. **5**, 053404 (2015)

14. Dai, G.L., Shang, J., Huang, J.P.: Theory of transformation thermal convection for creeping flow in porous media: cloaking, concentrating, and camouflage. Phys. Rev. E **97**, 022129 (2018)
15. Dai, G.L., Huang, J.P.: A transient regime for transforming thermal convection: cloaking, concentrating, and rotating creeping flow and heat flux. J. Appl. Phys. **124**, 235103 (2018)
16. Li, Y., Peng, Y.G., Han, L., Miri, M.A., Li, W., Xiao, M., Zhu, X.F., Zhao, J.L., Alu, A., Fan, S.H., Qiu, C.W.: Anti-parity-time symmetry in diffusive systems. Science **364**, 170–173 (2019)
17. Raman, A.P., Anoma, M.A., Zhu, L.X., Rephaeli, E., Fan, S.H.: Passive radiative cooling below ambient air temperature under direct sunlight. Nature **515**, 540–544 (2014)
18. Zhai, Y., Ma, Y.G., David, S.N., Zhao, D.L., Lou, R.N., Tan, G., Yang, R.G., Yin, X.B.: Scalable-manufactured randomized glass-polymer hybrid metamaterial for daytime radiative cooling. Science **355**, 1062–1066 (2017)
19. Chen, Z., Zhu, Z.X., Li, W., Fan, S.H.: Simultaneously and synergistically harvest energy from the sun and outer space. Joule **3**, 1 (2018)
20. Kubytskyi, V., Biehs, S.-A., Ben-Abdallah, P.: Radiative bistability and thermal memory. Phys. Rev. Lett. **113**, 074301 (2014)
21. Ordonez-Miranda, J., Ezzahri, Y., Tiburcio-Moreno, J.A., Joulain, K., Drevillon, J.: Radiative thermal memristor. Phys. Rev. Lett. **123**, 025901 (2019)
22. Milton, G.W.: The Theory of Composites. Cambridge University Press (2004)
23. Xu, L.J., Wang, R.Z., Huang, J.P.: Camouflage thermotics: a cavity without disturbing heat signatures outside. J. Appl. Phys. **123**, 245111 (2018)
24. Li, B.W., Wang, L., Casati, G.: Thermal diode: rectification of heat flux. Phys. Rev. Lett. **93**, 184301 (2004)
25. Li, B.W., Wang, L., Casati, G.: Negative differential thermal resistance and thermal transistor. Appl. Phys. Lett. **88**, 143501 (2006)
26. Wang, L., Li, B.W.: Thermal logic gates: computation with phonons. Phys. Rev. Lett. **99**, 177208 (2007)
27. Li, N.B., Ren, J., Wang, L., Zhang, G., Hänggi, P., Li, B.W.: Phononics: manipulating heat flow with electronic analogs and beyond. Rev. Mod. Phys. **84**, 1045–1066 (2012)
28. Bao, H., Chen, J., Gu, X.K., Cao, B.Y.: A review of simulation methods in micro/nanoscale heat conduction. ES Energy Environ. **1**, 16 (2018)
29. Ben-Abdallah, P., Biehs, S.-A.: Near-field thermal transistor. Phys. Rev. Lett. **112**, 044301 (2014)
30. Fernandez-Hurtado, V., Garcia-Vidal, F.J., Fan, S.F., Cuevas, J.C.: Enhancing near-field radiative heat transfer with Si-based metasurfaces. Phys. Rev. Lett. **118**, 203901 (2017)
31. Ghashami, M., Geng, H.Y., Kim, T., Iacopino, N., Cho, S.K., Park, K.: Precision measurement of phonon-polaritonic near-field energy transfer between macroscale planar structures under large thermal gradients. Phys. Rev. Lett. **120**, 175901 (2018)
32. Papadakis, G.T., Zhao, B., Buddhiraju, S., Fan, S.H.: Gate-tunable near-field heat transfer. ACS Photonics **6**, 709 (2019)
33. Hu, R., Zhou, S.L., Li, Y., Lei, D.Y., Luo, X.B., Qiu, C.W.: Illusion thermotics. Adv. Mater. **30**, 1707237 (2018)
34. Qu, Y.R., Li, Q., Cai, L., Pan, M.Y., Ghosh, P., Du, K.K., Qiu, M.: Thermal camouflage based on the phasechanging material GST. Light-Sci. Appl. **7**, 26 (2018)
35. Li, Y., Bai, X., Yang, T.Z., Luo, H., Qiu, C.W.: Structured thermal surface for radiative camouflage. Nat. Commun. **9**, 273 (2018)
36. Xu, L.J., Yang, S., Huang, J.P.: Thermal theory for heterogeneously architected structure: fundamentals and application. Phys. Rev. E **98**, 052128 (2018)
37. Xu, L.J., Yang, S., Huang, J.P.: Designing the effective thermal conductivity of materials of core-shell structure: theory and simulation. Phys. Rev. E **99**, 022107 (2019)
38. Xu, L.J., Yang, S., Huang, J.P.: Passive metashells with adaptive thermal conductivities: Chameleonlike behavior and its origin. Phys. Rev. Appl. **11**, 054071 (2019)
39. Xu, L.J., Huang, J.P.: Metamaterials for manipulating thermal radiation: transparency, cloak, and expander. Phys. Rev. Appl. **12**, 044048 (2019)

Chapter 20
Summary and Outlook

Abstract A summary of the book is presented. In particular, several key open questions are raised, which may help to encourage the reader to consider how to develop the field efficiently.

Keywords Theoretical thermotics · Transformation thermotics · Extended theories · Thermal metamaterials · The Stefan–Boltzman law · Nonlinear thermotics

20.1 Summary

In this book, we have presented eighteen theories (transformation thermotics and extended theories) for designing thermal metamaterials. The main research progresses have been introduced accordingly. In general, the concept of thermal metamaterials can help to design plenty of functional materials or devices with novel functions, which have broad potential applications in thermal protection, detection, treatment, and management.

20.2 Outlook: Future Directions and Open Questions

It is known that heat transfer has three basic ways: conduction, convection and radiation. So far, the role of thermal metamaterials in convection is still to be explored because of the difficulty in controlling mass diffusion at will. Moreover, even though the transformation method has been extended to treat thermal radiation (see Chaps. 4 and 5), we have only considered the Rosseland diffusion approximation [1], derived from the Stefan–Boltzman law, to deal with thermal radiation directly, which supposes that the mean free path of photons is far smaller than the system size. Therefore, the propagation of thermal radiation can be regarded as the diffusion of photons. In fact, there are many models to handle thermal radiation [2], and the Rosseland diffusion approximation is only one of them. The present establishment of

© Springer Nature Singapore Pte Ltd. 2020
J.-P. Huang, *Theoretical Thermotics*,
https://doi.org/10.1007/978-981-15-2301-4_20

transformation thermotics including radiation (Chaps. 4 and 5) benefits from the diffusive behavior of photons in optically thick media. Therefore, thermal radiation in Chaps. 4 and 5 is essentially a far-field effect. Certainly, it is also promising to extend the transformation theory into other radiative models, such as those considering near-field effects [3–6].

Although thermal metamaterials have achieved great attention within the past decade, several key scientific problems remain to be solved. For example, how to control thermal conduction, convection and radiation simultaneously by designing certain thermal metamaterials? Clearly, the existing progress is far from being satisfactory. Also, how to completely overcome the three limitations associated with transformation thermotics (anisotropic, inhomogeneous, and singular) is still an open question. In addition, how to freely tune thermal conductivities via thermal metamaterials, e.g., based on locally resonant nano-structures or periodic nano-structures, is still faced with challenges. Answers to these problems may promote future researches on theoretical thermotics and thermal metamaterials.

Furthermore, in the light of nonlinear optics, can one develop its counterpart in thermotics, nonlinear thermotics [7, 8]? Fundamentally, electric permittivities are a function of electric fields in nonlinear optics, and thermal conductivities are a function of temperatures in nonlinear thermotics. Since electric fields (or electric potentials) in nonlinear optics are only mathematically analogous to temperature gradients (or temperatures) in nonlinear thermotics, new physics and applications may be expected in this direction.

Last but not least, all the heat-conduction-related researches summarized in this book can be readily extended to other diffusion fields in electrostatics, magnetostatics, and particle diffusion, where electric conductivities, magnetic permeabilities and diffusion constants respectively play the same role as thermal conductivities in heat conduction.

References

1. Rosseland, S.: Theoretical Astrophysics. Oxford University Press, Clarendon, London and New York (1936)
2. Howell, J.R., Menguc, M.P., Siegel, R.: Thermal Radiation Heat Transfer, 6th edn. CRC Press, Boca Raton, London, New York (2016)
3. Ben-Abdallah, P., Biehs, S.A.: Near-field thermal transistor. Phys. Rev. Lett. **112**, 044301 (2014)
4. Fernández-Hurtado, V., Garcia-Vidal, F.J., Fan, S.H., Cuevas, J.C.: Enhancing near-field radiative heat transfer with Si-based metasurfaces. Phys. Rev. Lett. **118**, 203901 (2017)
5. Ghashami, M., Geng, H.Y., Kim, T., Iacopino, N., Cho, S.K., Park, K.: Precision measurement of phonon-polaritonic near-field energy transfer between macroscale planar structures under large thermal gradients. Phys. Rev. Lett. **120**, 175901 (2018)
6. Papadakis, G.T., Zhao, B., Buddhiraju, S., Fan, S.H.: Gate-tunable near-field heat transfer. ACS Photonics **6**, 709–719 (2019)
7. Dai, G.L., Shang, J., Wang, R.Z., Huang, J.P.: Nonlinear thermotics: nonlinearity enhancement and harmonic generation in thermal metasurfaces. Eur. Phys. J. B **91**, 59 (2018)
8. Yang, S., Xu, L.J., Huang, J.P.: Metathermotics: nonlinear thermal responses of core-shell metamaterials. Phys. Rev. E **99**, 042144 (2019)

Appendix
Brief History of the First Ten Years of Thermal Metamaterials

Abstract This appendix is an English version of the invitation article for celebrating the first ten years (2008–2018) of thermal metamaterials, whose original Chinese version was published in Physics, a Chinese counterpart of Physics Today, namely "J. P. Huang, Physics **47**, 685–694 (2018)". In this appendix, I introduce the brief history of thermal metamaterials for the first ten years. For this purpose, I choose 24 articles according to the chronological order of publication. I present the novel physics and new applications in the field. I expect this appendix could help the reader to quickly understand the whole field as well as its developing trend.

Keywords Thermal metamaterial · Thermal cloak · Thermal concentrator · Thermal rotator · Thermocrystal · Thermal diode · Energy-free thermostat · Daily radiation cooling

A.1   Opening Remarks

It has already been an accepted fact that 2008 is the first year of thermal metamaterials.

The field of thermal metamaterials is ten years old by 2018, if she is a child. According to Chinese customs, this is a year to celebrate. Therefore, I am glad to accept the invitation from "Physics", and write the history of the past ten years for her. The review has two main purposes: one is to commemorate the past, and the other is to prospect for the future. Throughout China, there are two genres in writing history: one is chronology, which is based on the happening time of historical events, such as "Spring and Autumn"; and the other is biography, which is based on a series of biographies, such as "Records of History". Compared with biography, chronology helps to present the order of different events and provide references for readers to rethink and innovate independently, although it seems to be slightly scrappy. In view of this, now I choose the chronological narrative mode to write the history for thermal metamaterials.

© Springer Nature Singapore Pte Ltd. 2020
J.-P. Huang, *Theoretical Thermotics*,
https://doi.org/10.1007/978-981-15-2301-4

In fact, the so-called metamaterials are a kind of artificially designed structural materials. Because of their special structures, metamaterials can possess some properties that do not exist in ordinary materials. Metamaterials (for optics/electromagnetism) can be traced back to the 1960s, but the rapid development was in the 1990s. Metamaterials can demonstrate new phenomena such as optical negative refraction. Since the 21st century, metamaterials have made great progress in the fields of acoustics, elastic wave, seismic wave, and mechanics, which is mainly due to the huge application demands in these fields. Because the aforementioned areas are mainly dominated by wave equations, their basic theories are similar. Differently, the theme of this appendix is thermal metamaterials (metamaterials in thermotics), which should be mainly described by diffusion equations. Thermal metamaterials give researchers a powerful method for manipulating macroscopic heat transfer. Over the past decade, hundreds of articles have been published. Considering the limited space, I would have to choose the following 24 research or review articles for the reader. In what follows, I will introduce these articles one by one in chronological order, in order to show the general situation of thermal metamaterials.

A.2   The Brief History of Ten Years: 2008–2018

(1) June 24, 2008: There is a white belly in the east, and the theory of thermal cloak firstly appeared.
Reference 1: C. Z. Fan, Y. Gao, and J. P. Huang, "Shaped graded materials with an apparent negative thermal conductivity", Applied Physics Letters 92, 251907 (2008).

Invisibility cloaks have always been a dream of human beings. In 2006, Pendry et al. and Leonhardt put forward the theory of electromagnetic cloak (wave systems), respectively. In 2008, Fan et al. (Ref. 1) firstly introduced the concept of electromagnetic cloak into the field of thermal conduction (diffusion systems). By referring to the theory of transformation optics, they established the theory of transformation thermotics and proposed the concept of thermal cloak, which can protect internal objects from thermal interference and does not disturb the temperature distribution outside the cloak, as if the internal objects do not exist. They further predicted the reverse of heat flux in the system, which is quite different from the traditional view that heat transfers from the region with high temperature to that with low temperature. Then they proposed the concept of apparent negative thermal conductivity based on the phenomenon. The relevant research results have potential applications in thermal protection, infrared deception, accurate temperature control, etc. Ref. 1 is the first article in the field of transformational thermotics. It had attracted little research interest in the first four years since its publication.

(2) September 15, 2008: The road is stepped out, and the second article of thermal cloak appeared.
Reference 2: T. Y. Chen, C.-N. Weng, and J.-S. Chen, "Cloak for curvilinearly anisotropic media in conduction", Applied Physics Letters 93, 114103 (2008).

Chen et al. (Ref. 2) also used the method of coordinate transformation to realize the thermal cloak. The difference is that Fan et al. (Ref. 1) studied isotropic background materials, while Chen et al. discussed the possibility of realizing thermal cloaks in anisotropic background materials. Their research showed that coordinate transformation theory is still applicable in this case. This work broadens the application of thermal cloak, and is of particular significance for specific nonuniform background.

(3) October 1, 2010: The third article of thermal cloak put forward the theory to design a bifunctional cloak in both thermal and electric fields.
Reference 3: J. Y. Li, Y. Gao, and J. P. Huang, "A bifunctional cloak using transformation media", Journal of Applied Physics 108, 074504 (2010).

With the development of transformation thermotics, the existing thermal cloak cannot satisfy the curiosity of researchers. Li et al. (Ref. 3) proposed a coordinate transformation to design a bifunctional cloak. Except for the function of thermal cloak, the bifunctional cloak can also act as an electric cloak. Bifunctional cloak provides a different idea for the miniaturization of bifunctional devices.

(4) March 2011: The fourth article on theoretical research of thermal cloak.
Reference 4: G. X. Yu, Y. F. Lin, G. Q. Zhang, Z. Yu, L. L. Yu, and J. Su, "Design of square-shaped heat flux cloaks and concentrators using method of coordinate transformation", Frontiers of Physics 6, 70 (2011).

With the development of thermal cloaks, other methods to control heat flux were explored intensively. In Ref. 4, Yu et al. proposed a method to design a square thermal cloak and thermal concentrator by using coordinate transformation theory. It is worth mentioning that thermal concentrator was theoretically predicted for the first time in this article. Most of the existing researches on thermal devices are based on the circular or spherical structure, so Ref. 4 provided the possibility to design arbitrarily shaped thermal devices.

(5) March 26, 2012: The fire becomes more blazing with more woods, and the fifth article on theoretical research of thermal cloak appeared.
Reference 5: S. Guenneau, C. Amra, and D. Veynante, "Transformation thermodynamics: cloaking and concentrating heat flux", Optics Express 20, 8207 (2012).

Fan et al. (Ref. 1) put forward the theory of transformation thermotics which is applicable to the steady-state thermal conduction equation. Guenneau et al. (Ref. 5) firstly proposed the unsteady-state transformation thermotics, and proved the reliability of the unsteady-state transformation thermotics based on the analytical theory and finite-element simulations. Importantly, they successfully designed transient thermal cloaks and thermal concentrators, and proposed a method to prepare thermal cloaks by using multilayer homogeneous structure instead of anisotropic materials, which provides theoretical guidance to experimental fabrication and practical application.

(6) May 21, 2012: Let the flame burn more fiercely, and the sixth article on thermal cloak, also the first experimental article, came to appear.
Reference 6: S. Narayana and Y. Sato, "Heat flux manipulation with engineered thermal materials", Physical Review Letters 108, 214303 (2012). Selected for Editors' suggestion.

In this article, Fan et al.'s (Ref. 1) theoretical prediction of both thermal cloak and apparent negative thermal conductivity was verified experimentally for the first time. This article and Ref. 5 were published in 2012, and both of them immediately aroused researches on thermal cloaks. It is worth mentioning that the previous Refs. 1–4 did not attract enough attention when they were published. Since 2012, they have also received widespread attention, especially Refs. 1–3.

Because of the strict limitation of transformation thermotics on the nonuniformity and extreme anisotropy, this kind of materials barely exists in nature, which makes the thermal cloak and other novel thermal devices face great challenges in experimental verification. In 2012, Narayana and Sato (Ref. 6) experimentally designed thermal cloak and other thermal devices for the first time. According to the effective medium theory, the radial and tangential thermal conductivities require different values. That is, the radial thermal conductivity can be regarded as the series connection of two kinds of thermal conductivities, while the tangential can be seen as the parallel connection of two kinds of thermal conductivities, so as to realize the different thermal conductivities in different directions and achieve the equivalent anisotropic thermal conductivity. Based on this method, they successfully overcome the requirement of anisotropy in thermal conductivities. By using the uniform isotropic materials, they can design thermal cloaks, thermal concentrators, and thermal rotators. Their work has opened a way for the control and realization of macroscopic heat transfer.

It is worth mentioning that after Ref. 6 was published, the Science magazine published a piece of news for it (http://www.sciencemag.org/news/2012/05/heat-trickery-paves-way-thermal-computers). The title is "Heat trickery pays way for thermal computers", which also contains some of my viewpoints in the field. In the news, I made some prospects for the future development of this field.

(7) June 21, 2012: Commentary on Refs. 1 and 6

Reference 7: P. Ball, "Against the flow", Nature Materials 11, 566 (2012).

In this review, Ball introduced the experimental verification (Ref. 6) and the theoretical predictions (Ref. 1) in detail, including thermal cloaking and apparent negative thermal conductivity. The experimental verification of the thermal concentrator was also commented in this review. More importantly, the compatibility between the apparent abnormal phenomenon and the laws of physics was discussed in the end of this review, which is helpful to eliminate doubts and promote the development of thermal metamaterials.

(8) January 11, 2013: A monument: controlling thermal phonons by using band gaps of periodic structure

Reference 8: M. Maldovan, "Narrow low-frequency spectrum and heat management by thermocrystals", Physical Review Letters 110, 025902 (2013).

Inspired by the modulation of electromagnetic waves by using photonic crystals, Maldovan proposed the concept of thermal crystals firstly, which can be used to regulate heat transfer arbitrarily. By using the coherent reflection of phonons, he calculated the band gap of the thermal crystal and found that the transport process of the thermal phonons can be precisely controlled at the microscopic level. He further theoretically designed the periodic structure in the nanoscale and calculated the corresponding band structure. Based on the theory, he proposed that such nanostructures could be used to design thermal devices or systems such as thermal waveguides, thermal imaging, thermal diodes, thermal clocks, and thermal super-lattices. In this work, the concept of thermal crystal was first proposed theoretically. This work also inspired many new researches.

(9) May 10, 2013: Experimental verification of transient thermal cloak

Reference 9: R. Schittny, M. Kadic, S. Guenneau, and M. Wegener, "Experiments on transformation thermodynamics: molding the Flow of Heat", Physical Review Letters 110, 195901 (2013). Selected for Editors' suggestion.

In this article, ten concentric rings with alternating thermal conductivities were used to realize the anisotropic thermal conductivity. The tangential conductivity is much larger than the radial one, and polydimethylsiloxane was used to fill the holes within copper. This structure can help to realize the thermal conductivity which is hard to achieve by using conventional materials, so that this structure could conform to the spatial distribution of thermal conductivity derived by coordinate transformation. This work experimentally verified the transient thermal cloak for the first time. It is worth mentioning that such thermal cloaks can be used to protect sensitive areas in circuits or chips from overheating damage in industry.

(10) June 27, 2013: The latest development in the field of thermal cloak was reviewed

Reference 10: U. Leonhardt, "Cloaking of heat", Nature 498, 440 (2013).

Professor Leonhardt is one of the pioneers in the field of transformation optics. The principle of optical cloak proposed by him and J. B. Pendry et al. was selected as the top ten scientific breakthroughs in 2006 by the Science magazine. He (Ref. 10) published a review article on the Nature magazine in 2013 which introduced the latest progress in the field of thermal cloak. This article plays an important role in promoting the development of thermal cloaks and thermal metamaterials.

(11) November 14, 2013: The original connotation of the name "thermal metamaterial"

Reference 11: M. Maldovan, "Sound and heat revolutions in phononics", Nature 503, 209 (2013).

Maldovan's article (Ref. 11) gave the field a formal name, namely, thermal metamaterial. What's more, the physical connotation contained in the name has a clear description, rather than a general summary. There is a special chapter in this article, whose title is "thermal metamaterials and the heat cloaking". In the chapter, for Refs. 1, 2, 5, 6 and 8, the author used "thermal metamaterial" to name the functions or devices such as thermal cloaks and other thermal devices which were designed by using the transformation thermotics. This article contributed to the formation of the direction "transformation thermotics and thermal metamaterial", and made the name "thermal metamaterial" widely recognized. In the article, the author gave a pertinent assessment of the existing research in this field, and described the researches as "innovative theoretical concepts", "exciting new technologies", "unprecedented control of heat flux", and "revolutionary developments".

Here, it must be noted that there is no doubt that Maldovan's review article (Ref. 11) plays an irreplaceable role in promoting the development of thermal metamaterial after 2013, and this article has also become a recognized source of the name of thermal metamaterial. However, in fact, the phrase "thermal metamaterial" first appeared in a collection of conference proceedings, which corresponded to the annual meeting of the American Society for Experimental Mechanics held in Indianapolis from June 7 to 10, 2010. The conference proceedings of the annual meeting include a paper with the title of "Thermal management and metamaterials" written by C. T. Roman, R. A. coutu and J. L. A. Starman. In the paper, the first two words in the abstract are just "thermal metamaterials". Unfortunately this paper did not attract great influences in the academic community as Ref. 11 did. So, it is a pity that this paper did not play a due role in promoting the development of thermal metamaterials.

(12) November 22, 2013: The latest development in thermal cloak was reviewed

Reference 12: M. Wegener, "Metamaterials beyond optics", Science 342, 939 (2013).

In this review, Wegener introduced the latest development of thermal cloak, and pointed out that the thermal diffusion length in the process of heat conduction can be compared with the optical wavelength (the characteristic length of the optical metamaterial). The square of thermal diffusion length is equal to the product of the thermal diffusion coefficient and the thermal diffusion time. Moreover, in this review, he pointed out that the negative thermal conductivity in the design of external thermal cloaks cannot exist because of the limitation of the second law of thermodynamics. But in active systems, the negative thermal conductivity can exist and does not violate the second law of thermodynamics because of the existence of heat source or cold source. These discussions on negative thermal conductivity provided a theoretical idea for the later design of thermal metamaterials. It is worth mentioning that his discussions on negative thermal conductivity are consistent with the discussions at the end of the article "EPL 104, 44001 (2013)".

(13) February 3, 2014: Theoretical design and experimental verification of the first three-dimensional thermal Cloak

Reference 13: H. Y. Xu, X. H. Shi, F. Gao, H. D. Sun, and B. L. Zhang, "Ultra-thin three-dimensional thermal cloak", Physical Review Letters 112, 054301 (2014). Selected for Editors' suggestion and a Viewpoint in Physics.

Xu et al. (Ref. 13) designed the first three-dimensional thermal cloak. The previous experiments of thermal clocks are two-dimensional devices, and three-dimensional devices could have extensive applicability undoubtedly. They successfully fabricated a three-dimensional ultra-thin thermal cloak by using sophisticated three-dimensional metal processing technology, where the thickness of copper is $100\,\mu m$ and the internal area radius is 0.5 cm.

(14) February 3, 2014: Theoretical design and experimental verification of a bilayer thermal cloak

Reference 14: T. C. Han, X. Bai, D. L. Gao, J. T. L. Thong, B. W. Li, and C.-W. Qiu, "Experimental demonstration of a bilayer thermal cloak", Physical Review Letters 112, 054302 (2014). Selected for Editors' suggestion and a Viewpoint in Physics.

The thermal cloaks based on the coordinate transformation theory generally require anisotropic thermal conductivities and extremely large range of thermal conductivities. Therefore, it is difficult to find corresponding materials in nature, which limits the preparation and application of thermal cloaks. To solve this problem, Han et al. (Ref. 14) designed a kind of bilayer thermal cloak. This cloak only needs two layers of uniform and isotropic materials. The thermal conductivities of the materials are calculated according to the Laplace equation. The temperature distributions in the cloak can be calculated by considering the associated boundary conditions, while the temperature distributions outside the cloak is not disturbed by the objects inside the cloak. This method works both for two-dimensional/three-dimensional cases and for steady/transient states. They have successfully fabricated a two-dimensional thermal cloak. Its inner (or outer) layer is made of polystyrene (or alloy) with a low (or high) thermal conductivity. The thickness of the inner and outer layers is 3.5 mm and 2.5 mm, respectively. The radius of the inner area is 6 mm. The background material of the cloak is thermal conductive silicone. Based on the sample, they measured the experimental data that are consistent with the theoretical prediction.

(15) March 2014: This article is the beginning of the research upsurge of thermal camouflage and thermal illusion
Reference 15: T. C. Han, X. Bai, J. T. L. Thong, B. W. Li, and C.-W. Qiu, "Full control and manipulation of heat signatures: cloaking, camouflage and thermal metamaterials", Advanced Materials 26, 1731 (2014).

On the basis of bilayer thermal cloaks, Ref. 15 designed a kind of thermal camouflage structure by directly solving the Laplace equation. The theory was verified by finite-element simulations and experiments. There are two semicircles on the left and right sides of the structure, which is the key to the design. The semicircles with low thermal conductivities help to insulate the heat flux. Therefore, whether there is an object inside the bilayer thermal cloak between the two semicircle structures or not, the temperature distribution in the background is the same. So the device can camouflage objects inside. This work also has potential applications in military and industry.

(16) November 12, 2014: The theoretical prediction of invisibility cloak with thermoelectric dual function is verified by experiments
Reference 16: Y. G. Ma, Y. C. Liu, M. Raza, Y. D. Wang, and S. L. He, "Experimental demonstration of a multiphysics cloak: manipulating heat flux and electric current simultaneously", Physical Review Letters 113, 205501 (2014).

In 2010, Li et al. (Ref. 3) theoretically designed a type of cloak that works for thermal and electric fields, respectively. Subsequently, Ma et al. (Ref. 16) experimentally fabricated such a cloak. They used the idea of bilayer thermal cloak for reference, and found that the bilayer cloak structure can have the dual functions as long as the ratio between background thermal conductivity and electric conductivity is equal to the ratio between outer layer thermal conductivity and electric conductivity. In the experiment, they abandoned the traditional metal-insulator composite structure and adopted semiconductor silicon to realize their idea. The outer layer of the thermal cloak is low-density doped n-type silicon, and the inner layer is air film, and the background material is silicon with periodic holes. The holes are filled with polydimethylsiloxane. According to the effective medium theory, the ratio of background thermal conductivity to electric conductivity can meet the requirement when the hole/silicon area fraction is appropriate. The experimental results verified their idea. Their work, together with the previous articles of Li et al. (Ref. 3) and Moccia et al. [Physical Review X 4, 021025 (2014)], successfully extended the cloak from the single physical (thermal) field to the multiphysical (thermal and electric) fields, and provided a reliable theoretical basis and experimental methods for subsequent researches.

(17) November 27, 2014: Another monument: the principle and prototype of daytime radiative cooling came out

Reference 17: A. P. Raman, M. A. Anoma, L. X. Zhu, E. Rephaeli, and S. H. Fan, "Passive radiative cooling below ambient air temperature under direct sunlight", Nature 515, 540 (2014).

The energy consumption of refrigeration system accounts for 15% of the global energy consumption [E. A. Goldstein, A. P. Raman, and S. H. Fan, Nature Energy 2, 17143 (2017)], and the proportion is even as high as 50% in the United States. Compared with the "active refrigeration" devices such as refrigerators and air conditioners that usually need energy input, a "passive refrigeration" device needs no energy input. So the passive refrigeration device can greatly reduce the proportion of refrigeration system in energy consumption, so as to effectively alleviate the energy crisis. Fortunately, as early as 1967, the concept of radiation cooling was proposed by F. Trombe, which was verified by experiments in 1975 [S. Catalanatti, V. Cuomo, G. Piro, D. ruggi, V. silvestrini, and G. Troise, "The radiation cooling of selective surfaces", Solar Energy 17, 83 (1975)]. However, the devices work only at night. A large number of researchers try to find an effective way to achieve passive cooling in the daytime, but there had been no substantial progress for a long time, and even some people predicted that it is impossible to achieve radiation cooling in the daytime.

However, the results reported by Ref. 17 successfully broke the above prediction. The final stable temperature of an object is determined by the absorbed heat minus the emitted heat. Except for the thermal conduction and convection between the refrigeration system and the environment, the input energy comes from the solar radiation (corresponding to the wavelength of $0.3–2.5$ μm), and the output energy is emitted by the thermal radiation (this part of energy can radiate the heat to the vast universe through the atmospheric window $8–13$ μm). By designing the absorption and emission spectrum of the material, the "absorptivity/emissivity" in the $0.3–2.5$ μm band range of the material can be as small as possible, while the "absorptivity/emissivity" in the $8–13$ μm band range can be as large as possible. Thus the "passive cooling" in the daytime can be realized. The authors (Ref. 17) used SiO_2, HfO_2, and other materials to fabricate the photonic-crystal thin film that meets the above theoretical requirements. Their results show that the surface temperature of the film can be 4.9 °C lower than the ambient temperature when it is exposed to direct sunlight in the daytime. This is the first time to realize "daytime passive cooling". If such technology can be applied for large scales (such as covering the top of residential houses with this kind of material), it can greatly reduce the dependence on air conditioning in summer. Definitely, this could reduce energy consumption significantly.

(18) July 17, 2015: Learn from nature how to design thermal metamaterials

Reference 18: N. Shi, C.-C. Tsai, F. Camino, G. D. Bernard, N. Yu, and R. Wehner, "Keeping cool: enhanced optical reflection and radiative heat dissipation in Saharan silver ants", Science 349, 298 (2015).

The method of designing photonic crystals in Ref. 17 needs to control the thickness and growth of each film layer, which is not convenient for large-scale production and applications. The silver ants in the Sahara Desert can maintain a temperature of 48–51 °C in summer when the ambient temperature is as high as 60–70 °C. Organisms usually do not have a large number of SiO_2, HfO_2 and other substances, not to mention the above-mentioned artificial photonic crystal structure, so the study of the heat dissipation mechanism of silver ant is helpful to develop a more effective radiation refrigeration mechanism. Through scanning electron microscope observation of the head of silver ant, Shi et al. (Ref. 18) found that there are a large number of tiny triangular hairs, which can reflect the solar radiation and emit the energy through atmospheric window, thus maintaining a relatively low temperature. This kind of microstructure matched with the wavelength of solar radiation can be completed without complicated photonic crystal structure design. The tiny triangular hair structures of the silver ant are of great significance to the excavation of new radiation refrigeration mechanism.

(19) November 5, 2015: Nonlinear transformation theory and macroscopic thermal diode

Reference 19: Y. Li, X. Y. Shen, Z. H. Wu, J. Y. Huang, Y. X. Chen, Y. S. Ni, and J. P. Huang, "Temperature-dependent transformation thermotics: from switchable thermal cloaks to macroscopic thermal diodes", Physical Review Letters 115, 195503 (2015). Cover Article.

Common materials used in the existing transformation thermotics are not enough to solve the problem of macroscopic thermal rectification by designing macroscopic thermal diodes. This is because the thermal conductivity of common materials is generally independent of temperature, while the thermal diodes need to transfer heat flux in one direction and prohibit heat flux in the opposite direction. Therefore, it is necessary to use nonlinear materials, that is, the thermal conductivity depends on temperature. In Ref. 19, the authors proposed a switchable thermal cloak based on the nonlinear materials, and obtained the selective response of the thermal cloak to the ambient temperature. Therefore, it is also called intelligent thermal cloak. At the same time, they also designed a kind of macroscopic thermal diode. This diode has potential applications, such as heat preservation, heat dissipation, and energy conservation. Moreover, if macroscopic thermal triodes can be designed based on it, the macroscopic thermal logic gate can also be expected, which provides a different idea for efficient utilization of heat energy.

(20) July 29, 2016: Energy-free maintenance of constant temperatures as ambient temperature gradients change.

Reference 20: X. Y. Shen, Y. Li, C. R. Jiang, and J. P. Huang, "Temperature trapping: energy-free maintenance of constant temperatures as ambient temperature gradients change", Physical Review Letters 117, 055501 (2016). Selected as Focus in APS Physics.

Heat energy is a kind of low-quality energy for most of the time, which cannot be fully utilized. If people can make full use of the heat energy in the environment, it would be greatly beneficial for energy conservation and environmental protection. Shen et al. (Ref. 20) proposed a temperature trapping theory by using phase change materials. The so-called phase change means that the thermal conductivity of the materials can respond to temperature. Then they designed a new type of thermostat, which can maintain the required constant temperature without the need of additional energy as ambient temperature changes. As an application of the concept, the authors designed a different thermal cloak. Although the ambient temperature gradient changes significantly, the temperature in the central area of the cloak is still a constant, which is significantly different from the existing thermal cloaks. Their work provides a different idea for controlling heat flux with energy saving.

(21) March 10, 2017: Breakthroughs in basic research and industrialization of daytime radiative cooling

Reference 21: Y. Zhai, Y. G. Ma, S. N. David, D. L. Zhao, R. N. Lou, G. Tan, R. G. Yang, and X. B. Yin, "Scalable-manufactured randomized glass-polymer hybrid metamaterial for daytime radiative cooling", Science 355, 1062 (2017).

The principle of daytime cooling refrigeration has been generally accepted, which is to reduce the absorption of solar radiation and increase the thermal radiation through the atmospheric window. However, how to achieve such a theoretical requirement? The main experimental methods rely on photonic crystal method (Ref. 17) and polymer photon method (Ref. 21). Among them, the polymer photon method is receiving more and more attention, which is mainly because it has the following advantages: the experimental process is relatively simple, the technical requirements are low, and the preparation cycle is short. The specific details of the method are as follows: the randomly distributed SiO_2 particles with a volume fraction of about 6% are doped in the polymer, so that the polymer can have a strong scattering effect with the wavelength of 8–13 μm, and the polymer is transparent with the wavelength of 0.3–2.5 μm. The sunlight through the polymer is reflected by the silver film on the back. In this way, the theoretical requirements are satisfied well. Based on the composite structure of organic polymer and silver film, Zhai et al. (Ref. 21) fabricated the samples which can reflect 96% of the solar radiation, and the emissivity in the range of atmospheric window reaches 0.93. Because their method does not have many restrictions on experimental process, the materials can be prepared rapidly and massively, which have been put into production and commercialized in reality.

(22) August 29-September 1, 2017: Thermal metamaterials begin to have practical applications in printed circuit boards

Reference 22: E. M. Dede, F. Zhou, P. Schmalenberg, and T. Nomura, "Thermal metamaterials for heat flow control in electronics", Proceedings of the ASME 2017 International Technical Conference and Exhibition on Packaging and Integration of Electronic and Photonic Microsystems, article No. IPACK2017-74112 (2017).

In 2014, Dede and his collaborators proposed a kind of thermal metamaterial with "thermal fiber" structure. In 2017, they (Ref. 22) applied this thermal fiber structure to the electronic components. By using theoretical calculation and experimental research, they found that the geometrically optimized copper fiber can be grown on the printed circuit board. This design can protect the thermal elements on the circuit board effectively. Compared with the circuit board without any copper fiber structure, the optimized structure can reduce the temperature of the thermistor by 10.5 °C, which is the first time to apply thermal metamaterial to the electronic circuits. This structure can also be used for heat collection, so as to improve the efficiency of thermoelectric conversion and heat dissipation.

(23) January 18, 2018: How to make objects emit false infrared thermal radiation signals

Reference 23: Y. Li, X. Bai, T. Yang, H. Luo, and C.-W. Qiu, "Structured thermal surface for radiative camouflage", Nature Communications 9, 273 (2018).

Based on the blackbody radiation theory, it is known that any object can radiate different wavelengths of electromagnetic waves according to different temperatures. Infrared imaging technology has been used in many fields, such as temperature measurement, night vision, and infrared guided missiles. Therefore, thermal radiation camouflage has high academic significance and application value. When the temperature of an object is different from that of the background or its thermal radiation is stronger or weaker than that of the background, the object can be detected by infrared imaging devices. Therefore, the current common practice to achieve thermal radiation camouflage is to cover the surface of the object with a carefully designed film, so as to adjust the thermal radiation from the object to match that from the background. For this approach, one has to know the background temperature and the properties of the object in advance, which limits the application. In contrast, Li et al. (Ref. 23) designed a special thermal metamaterial for thermal radiation camouflage. This kind of thermal metamaterial can be directly applied to the objects and backgrounds to change the properties of thermal radiation without the need to know the background temperature in advance. This method is convenient and useful for applications.

(24) May 29, 2018: To hide heat sources.

Reference 24. R. Hu, S. L. Zhou, Y. Li, D.-Y. Lei, X. B. Luo, and C.-W. Qiu, "Illusion thermotics", Advanced Materials 30, 1707237 (2018).

Thermal metamaterials make many novel thermal phenomena possible, and thermal illusion is one of them. In fact, any object in nature can be regarded as a heat source, because they all emit thermal radiation. Nevertheless, there is surprisingly little research on heat source illusion. In Ref. 24, Hu et al. proposed that a single heat source can be disguised into multiple virtual heat sources, and the observer cannot distinguish the actual heat source through the external temperature distribution. In other words, the location, shape, size, and quantity of the actual heat source are hidden. They (Ref. 24) developed a transformation location method based on the transformation thermotics [Q. W. Hou, X. P. Zhao, T. Meng, and C. L. Liu, Applied Physics Letters 109, 103506 (2016)]. They gave the analytical results, and verified the theory by conducting finite-element simulations and experiments. The illusion device reported in this article is simple and convenient for practical applications.

A.3   Outlook

The past is for a better look to the future. I hope that the introduction of the above 24 articles can help the reader to understand the general situation of the field, "theoretical thermotics and thermal metamaterials".

It can also be seen that the development of a field requires the joint efforts of many factors, such as the international cooperation among researchers, the analogy between different research methods, the cooperation between theory and experiment, and so on. I think the combination of basic researches and industrial applications would become a major development trend of thermal metamaterials. An excellent example is about daytime radiative cooling: Professor Ronggui Yang and his team from the University of Colorado have made outstanding achievements. In fact, researchers in this field have published many articles in the journals of engineering thermophysics, which shows that peers from the engineering field make great contributions to this field as well. On the other hand, the theoretical methods in this field need to be further developed and mined. For example, how to develop the corresponding theory of nonlinear optics into the field of thermotics based on the temperature-dependence of thermal conductivities? How to simultaneously control the three basic ways for heat transfer (thermal conduction, convection and radiation) by developing transformation thermotics? How to improve the utilization efficiency of energy materials (such as thermoelectric materials) through the unprecedented control of thermal metamaterials? To solve these problems, we need more young people who are willing to accept the challenge to join us. Of course, we also need the input and persistence of peers.

Writing the outlook is essentially a prediction of the future, and I hope this outlook could become the historical material of the future future.

Printed in the United States
By Bookmasters